Studies in Big Data

Volume 37

Series editor

Janusz Kacprzyk, Polish Academy of Sciences, Warsaw, Poland
e-mail: kacprzyk@ibspan.waw.pl

The series "Studies in Big Data" (SBD) publishes new developments and advances in the various areas of Big Data- quickly and with a high quality. The intent is to cover the theory, research, development, and applications of Big Data, as embedded in the fields of engineering, computer science, physics, economics and life sciences. The books of the series refer to the analysis and understanding of large, complex, and/or distributed data sets generated from recent digital sources coming from sensors or other physical instruments as well as simulations, crowd sourcing, social networks or other internet transactions, such as emails or video click streams and other. The series contains monographs, lecture notes and edited volumes in Big Data spanning the areas of computational intelligence incl. neural networks, evolutionary computation, soft computing, fuzzy systems, as well as artificial intelligence, data mining, modern statistics and Operations research, as well as self-organizing systems. Of particular value to both the contributors and the readership are the short publication timeframe and the world-wide distribution, which enable both wide and rapid dissemination of research output.

More information about this series at http://www.springer.com/series/11970

Artur Gramacki

Nonparametric Kernel Density Estimation and Its Computational Aspects

Artur Gramacki
Institute of Control and Computation
Engineering
University of Zielona Góra
Zielona Góra
Poland

ISSN 2197-6503 ISSN 2197-6511 (electronic)
Studies in Big Data
ISBN 978-3-319-89094-4 ISBN 978-3-319-71688-6 (eBook)
https://doi.org/10.1007/978-3-319-71688-6

Printed on acid-free paper

This Springer imprint is published by Springer Nature
The registered company is Springer International Publishing AG
The registered company address is: Gewerbestrasse 11, 6330 Cham, Switzerland

To my brother Jarek

Foreword

In the rapidly developing field of processing and analyzing big data, new challenges come not only from a vast volume of data but also from their smoothing. There are many smoothing techniques known, but it seems that one of the most important ones is kernel smoothing. It belongs to a general class of nonparametric estimation of functions such as probability density functions or regression ones. Kernel-based estimators of such functions are very attractive for practitioners as they can uncover important patterns in the data while filtering noise and ignoring irrelevant details. Subsequently, the estimated function is smooth, and the level of smoothness can be typically controlled by a parameter known as a bandwidth.

The subject matter of this book is primarily kernel probability density function estimation. During the last few decades, intensive research has been conducted in this area, and it seems that from a theoretical point of view, nonparametric kernel density estimation has reached its maturity. Meanwhile, relatively less research has been devoted to computational problems regarding kernel density estimation and optimal bandwidth selection. These problems are very important in the context of the need of analyzing big datasets, both uni- and multidimensional.

A part of the book focuses on fundamental issues related to nonparametric density estimation to give the readers intuition and basic mathematical skills required to understand kernel smoothing. This part of the book is of tutorial value and can be perceived as a good starting point for readers unfamiliar with non-parametric techniques. The book is also meant for a more advanced audience, interested in recent developments related to very fast and accurate kernel density estimation as well as bandwidth selection.

A unified framework based on the fast Fourier transform is presented. Abundant experimental results included in the book confirm its practical usability as well as very good performance and accuracy. Additionally, some original preliminary research on using modern FPGA chips for bandwidth selection is presented. All the concepts described in this book are richly illustrated with both academic examples and real datasets. Also, a special chapter is devoted to interesting and important examples of practical usage of kernel-based smoothing techniques.

Concluding, the book can be strongly recommended to researchers and practitioners, both new and experienced not only in the field of data smoothing but also in wide understanding of processing and analyzing big data.

Zielona Góra, Poland
September 2017

Józef Korbicz
Professor,
Corresponding member of the Polish
Academy of Sciences

Preface

This book concerns the problem of data smoothing. There are many smoothing techniques, yet the kernel smoothing seems to be one of the most important and widely used ones. In this book, I focus on a well-known technique called kernel density estimation (KDE), which is an example of a nonparametric approach to data analysis.

During the last few decades, many books and papers devoted to this broad field have been published, so it seems that this area of knowledge is quite well understood and reached its maturity point. However, many (or even most) of the practical algorithms and solutions designed in the context of KDE are very time-consuming with quadratic computational complexity being a commonplace. This might be not problematic for situations, where datasets are not that big (at the level of hundreds of individual data points) but it already can be an obstacle for datasets containing thousands or more individual data points, especially in case of multivariate data. Progress in terms of theoretical results related to KDE does not go hand in hand with the development of fast and accurate algorithm for speeding up the required calculations in practical terms. In this sense, this book can be considered a valuable contribution to the field of KDE.

This book is a result of my research in the area of numerical and computational problems related to KDE, an interest that has been developing since ca. 2010. It should be viewed primarily as a research monograph and is intended both for those new to such topics as well as for more experienced readers. The first few chapters present a background material, describing the fundamental concepts related to the nonparametric density estimation, kernel density estimation, and bandwidth selection methods. The presented material is richly illustrated by numerical examples, using both toy and real datasets. The following chapters are devoted to the presentation of our own research on fast computation of kernel density estimators and bandwidth selection. The developed methods are based on the fast Fourier transform (FFT) algorithm that relies on a preliminary data transformation known as data binning. Certain results obtained by me on utilizing field-programmable gate arrays (FPGA) in the context of fast bandwidth selection are also included. FPGA devices are a not so common choice in terms of implementing purely numerical algorithms.

The proposed implementation can be seen as a preliminary study of practical usability of such FPGA-based applications. The monograph ends with a chapter presenting a number of applications related to KDE. The following example applications are given: discriminant analysis, cluster analysis, kernel regression, multivariate statistical process control, and flow cytometry.

I wish to express my deepest gratitude to Prof. Józef Korbicz, Ph.D., D.Sc., a corresponding member of the Polish Academy of Sciences for his continuing support since 2000 to this day and for motivating me to work hard on the problems at hand. I would also like to express my sincere thanks to Prof. Andrzej Obuchowicz, Ph.D., D.Sc., for his help at a time of particular need in my life. I also extend my thanks to Prof. Dariusz Uciński, Ph.D., D.Sc., and to Marcin Mrugalski, Ph.D., D.Sc., for their continuing kindness. Marek Sawerwain, Ph.D., helped me a lot in all programming tasks related to FPGA, and I thank him very much for this. My warm thanks also go to my mother, to my wife Edyta, and to my children Ola and Kuba for their love, patience, and support. Finally, I would like to express my sincere gratitude to my brother Jarosław because I truly believe that writing this book would not have been possible without his powerful and continuous assistance and support, both in professional and in family life.

Zielona Góra, Poland Artur Gramacki
September 2017

About this Book

This book describes computational problems related to kernel density estimation (KDE)—one of the most important and widely used data smoothing techniques. A very detailed description of novel FFT-based algorithms for both KDE computations and bandwidth selection is presented.

The theory of KDE appears to have matured and is now well developed and understood. However, there is not much progress observed in terms of performance improvements. This book is an attempt to remedy this.

The book primarily addresses researchers and advanced graduate or postgraduate students who are interested in KDE and its computational aspects. The book contains both some background and much more sophisticated material, hence also more experienced researchers in the KDE area may find it interesting.

The presented material is richly illustrated with many numerical examples using both artificial and real datasets. Also, a number of practical applications related to KDE are presented.

Contents

About the Author

Artur Gramacki is an Assistant Professor at the Institute of Control and Computation Engineering of the University of Zielona Góra, Poland. His main interests cover general exploratory data analysis, while recently he has focused on parametric and nonparametric statistics as well as kernel density estimation, especially its computational aspects. In his career, he has also been involved in many projects related to the design and implementation of commercial database systems, mainly using Oracle RDBMS. He is a keen supporter of the R Project for Statistical Computing, which he tries to use both in his research and teaching activities.

Abbreviations

AMISE	Asymptotic MISE
AMSE	Asymptotic Mean Squared Error
ASH	Averaged Shifted Histogram
BCV	Biased Cross-Validation
CDF	Cumulative Distribution Function
CV	Cross-Validation
DA	Discriminant Analysis
FFT	Fast Fourier Transform
FPGA	Field-Programmable Gate Arrays
GPU	Graphics Processing Units
HDL	Hardware Description Languages
HLS	High-Level Synthesis
ISB	Integrated Squared Bias
ISE	Integrated Squared Error
KCDE	Kernel CDF Estimation (or Estimator depending of a context)
KDA	Kernel Discriminant Analysis
KDE	Kernel Density Estimation (or Estimator depending of a context)
KNN	K-Nearest Neighbors
KNR	Kernel Nonparametric Regression
LSCV	Least-Squares Cross-Validation
MIAE	Mean Integrated Absolute Error
MISE	Mean Integrated Squared Error
MPI	Message Passing Interface
MS	Maximal Smoothing
MSE	Mean Squared Error
NS	Normal Scale
PI	Plug-in

ROT	Rules-of-Thumb
RTL	Register-Transfer Level
SCV	Smoothed Cross-Validation
UCV	Unbiased Cross-Validation

Notation

$\otimes$	Kronecker product operator
$\odot$	Element-wise multiplication
$\mathbb{1}_{\{x \in A\}}$	Indicator function, that is $\mathbb{1}_{\{x \in A\}}= 1$ for $x \in A$ and $\mathbb{1}_{\{x \in A\}}= 0$ for $x \notin A$
$\mathbf{A}$	$d \times d$ matrix
$\boldsymbol{a}$	Vector of size d
c_i	Univariate binning grid counts
$\boldsymbol{c}_i$	Multivariate binning grid counts
$\mathsf{D}f$	First derivative (gradient) of f
$\mathsf{D}^{\otimes r}$	r-th Kronecker power of the operator $\mathbf{D}$
d	Problem dimensionality
F	Unknown cumulative distribution function
$\hat{F}$	Kernel cumulative distribution function estimate
$\mathcal{F}$	Fourier transform operator
$\mathcal{F}^{-1}$	Inverse Fourier transform operator
f	Unknown density function
$f^{(r)}$	Unknown density derivative function of order r
$\hat{f}^{(r)}$	Kernel density function estimate
$\tilde{f}$	Kernel density function estimate after binning employed
$\hat{f}^{(r)}$	Kernel density derivative function estimate of order r
$\hat{f}(x; h)$	Univariate kernel density estimate at point x and with bandwidth scalar h
$\hat{f}_{-i}(x; h)$	Univariate leave-one-out kernel density estimate at point x and with bandwidth scalar h
$\hat{f}(\boldsymbol{x}; \mathbf{H})$	Multivariate kernel density estimate at point $\boldsymbol{x}$ and with bandwidth matrix $\mathbf{H}$
$\hat{f}_{-i}(\boldsymbol{x}; \mathbf{H})$	Multivariate leave-one-out kernel density estimate at point x and with bandwidth matrix $\mathbf{H}$

$f_1 * f_2$	Convolution of functions f_1 and f_2, that is $f_1 * f_2(x) = \int f_1(u) f_2(x-u)\, du$
G	$d \times d$ pilot bandwidth matrix which is symmetric and positive definite
$\mathcal{G}$	Set of classes in Discriminant Analysis
g	Pilot bandwidth
g_i	Univariate binning grid points
$\boldsymbol{g}_i$	Multivariate binning grid points
$\mathsf{H}f$	Hessian operator of f
$\mathcal{H}_r$	r-th order Hermite polynomial
$\mathbf{H}$	$d \times d$ bandwidth matrix (smoothing matrix) which is symmetric and positive definite
h	Bandwidth scalar (smoothing parameter)
$\mathbf{H}_{\mathcal{F}}$	Class of symmetric, positive definite $d \times d$ matrices
$\mathbf{H}_{\mathcal{D}}$	Subclass of diagonal positive definite $d \times d$ matrices
$\mathbf{H}_{\mathcal{S}}$	Subclass of a positive constant times the identity matrix
I	Antiderivative of the kernel function K
K	Unscaled kernel function
$\mathcal{K}$	Symmetric univariate kernel, this symbol was used in Sect. 3.3.2 to differentiate univariate ($\mathcal{K}$) and multivariate ($\mathcal{K}$) kernels
K^{P}	Product kernel
K^{R}	Radially symmetric kernel
$K * K$	Convolution of kernel K with itself, that is $K * K(x) = \int K(u) K(x-u)\, du$
$K_{\mathbf{H}}$	Scaled kernel function with bandwidth matrix $\mathbf{H}$
K_h	Scaled kernel function with bandwidth scalar h
L	Pilot kernel (the first meaning of the symbol)
L	Values which is $L<M-1$ (the second meaning of the symbol)
M	Number of grid points
$\boldsymbol{m}_{\mathbf{H}}(\boldsymbol{x})$	Mean shift operator
$\mathcal{N}(x; \mu, \sigma)$	Univariate normal distribution with mean μ and standard deviation σ at point x
$\mathcal{N}(\boldsymbol{x}; \boldsymbol{\mu}, \boldsymbol{\Sigma})$	Multivariate normal distribution with mean vector $\boldsymbol{\mu}$ and covariance matrix Σ at vector point $\boldsymbol{x}$
n	Random sample size (usually size of experimental data)
$R(g) = \int g(x)^2\, dx$	(g is a real-valued univariate function)
T	Geometric mean of the values of KDE of all data points, that is $\hat{f}(x_1, h), \hat{f}(x_2, h), \ldots, \hat{f}(x_n, h)$
V	Volume of the region $\mathcal{R}$
vec$\mathbf{A}$	The vectorization operator, it is a transformation that converts a matrix $\mathbf{A}$ into a column vector by stacking the columns of this matrix on top of one another

vech**A**	The half-vectorization operator (or vector half operator), it is defined only for a symmetric matrix **A** by vectorizing only the lower triangular part of **A**
X	the univariate data vector
$X_1, X_2, \ldots, X_n$	Univariate random sample of size n (usually experimental data)
$\boldsymbol{X}_1, \boldsymbol{X}_2, \ldots, \boldsymbol{X}_n$	Multivariate random sample of size n and dimension d (usually experimental data). $\boldsymbol{X}_i = (X_{i1}, X_{i2}, \ldots, X_{id})^T, i = 1, 2, \ldots, n$
x	Univariate data point
$\lceil x \rceil$	Ceiling(x), least integer greater than or equal to x
$\lfloor x \rfloor$	Floor(x), greatest integer less than or equal to x
δ_k	Mesh size (used in binning)
λ	Largest eigenvalue of **H**
$\mu_l(g) = \int x^l g(x)\, dx$	(l-th central moment of g, g is a real-valued univariate function)
π_k	Prior probability of the class k in Discriminant Analysis
$\hat{\pi}_k$	Sample proportion of the class k in Discriminant Analysis
$\phi(x)$	Density of the standard univariate normal distribution having mean zero and variance equal to one
$\phi_\sigma(x)$	Density of the univariate normal distribution having mean zero and variance σ
$\phi_{\boldsymbol{\Sigma}}(\boldsymbol{x})$	Density of the multivariate normal distribution having mean zero and covariance matrix Σ
Ψ_r	Integrated density derivative functional Ψ_r $= \int f^{(r)}(x) f(x)\, dx($ r an even integer, $\Psi_r = 0$ if r is odd)
$\boldsymbol{\Psi_4}$	Matrix of fourth-order density derivative functionals of dimension $\frac{1}{2}d(d+1) \times \frac{1}{2}d(d+1)$
$\boldsymbol{\psi_4}$	Vector of fourth-order density derivative functionals of dimension d^4

List of Figures

List of Tables

parametric methods try to attempt an estimation of the density directly from the data, without making any parametric assumptions about the underlying distribution.

There exists a number of methods for nonparametric density estimation, based on e.g. kernel smoothing, histograms, orthogonal series, splines, frequency polygons, wavelets or the penalized likelihood. An extensive survey of these topics can be found for example in [42, 97, 134, 179]. Our method of choice is kernel density estimation, given that it can be easily interpreted and is very often used in practical applications. Moreover, KDE has recently been the subject of an extensive research, making it a matured tool to be used in data analysis. The literature on nonparametric density estimation is huge and we cannot list even roughly complete list of articles or books on the subject. The good starting books on the subject are [82, 158, 168, 170, 188].

Most nonparametric methods are very time consuming, with computational complexity of $O(n^2)$ (or greater) being a common result in most cases—in that respect, KDE is not an exception. In addition, a typical KDE requires a pre-estimate of a parameter known as a *bandwidth* or a *smoothing parameter* and this procedure is also very time consuming (except for some simplified and often not too accurate *rule-of-thumb* methods). There has been relatively little research devoted to the problem of fast KDE and bandwidth computations. We describe this problem in some detail in the context of the FFT-based and FPGA-based approaches. Note that there exists yet another approach, based on a technique know as *Fast Gauss Transform* [74], that has also been studied (see [111, 142–144]). However, despite certain interesting features of this technique, it is not covered in this book.

It should be also stressed that KDE is of practical importance only in terms of low data dimensionality with six-dimensional data considered the upper limit. In higher dimensions, the sparsity of data leads to very poor and unstable estimators, pointing towards the use of other techniques. Alternatively, a preliminary data dimensionality reduction could also be performed (for the up-to-date techniques, see for example [174]). This phenomenon is often called *the curse of dimensionality* (see Sect. 3.9).

1.2 Contents

Chapters 2–4 provide an overview of the background for nonparametric density estimation. Chapters 5 and 6 present the author's contribution to KDE research area. Chapter 7 shows some selected applications related to KDE. The presented material is richly illustrated with numerical examples, using both toy and real datasets.

We now describe the contents of each chapter in more detail.

Nonparametric density estimation. Chapter 2 presents a background material, describing the fundamental concepts related to the nonparametric density estimation. First, a well-known *histogram* technique is briefly presented together with a description of its main drawbacks. To avoid the highlighted problems, at least to some extent, one might use a smart histogram modification known in the literature as an *averaged shifted histogram* (ASH). A simple example presented in this chapter shows its advantages over the classical histogram. The subsequent part of the chapter gives a general formulation of nonparametric density estimation, followed by the presentation of two basic approaches (the *Parzen windows* and the *k-nearest neighbors* (KNN)). Finally, the main advantages and drawbacks of the presented methods are described. The chapter is richly illustrated with a number of numerical examples.

Kernel density estimation. Chapter 3 describes the kernel density estimation technique that can be considered a smoothed version of the Parzen windows presented in the previous chapter. First, the most popular kernel types are presented together with a number of basic definitions both for uni- and multivariate cases and then a review of performance criteria is provided, starting with the univariate case and then extended to the general multivariate case. The subsequent part of the chapter is devoted to an introduction of two important KDE extensions, namely *adaptive* KDE and KDE with *boundary correction*. The notion of *kernel density derivative estimation* (KDDE) is also presented. The final part of the chapter describes how KDE can be used for nonparametric estimation of *cumulative distribution function* (CDF). The chapter ends with some notes on computational aspects related to KDE.

Bandwidth selectors for kernel density estimation. Chapter 4 describes the most popular bandwidth selection methods (also known as *bandwidth selectors*). It starts with a description of the constrained and unconstrained bandwidth selectors and moves on to an overview of the three major types of selectors (that is: *rule-of-thumb* (ROT), *cross-validation* (CV) and *plug-in* (PI) selectors). The next part of the chapter is devoted to describing these selectors in more detail, both for the uni- and multivariate cases. Finally, a few numerical examples are given. The chapter is rounded off with a short section on the computational issues related to bandwidth selectors.

FFT-based algorithms for kernel density estimation and bandwidth selection. Chapter 5 is devoted to presentation of the results of author's own research on fast computation of kernel density estimators and bandwidth selection. The developed methods are based on the *fast Fourier transform* (FFT) algorithm that relies on a preliminary data transformation known as *data binning*. We start with the description of binning rules, both for the univariate and multivariate cases, and provide the reader with several numerical examples. The main aim of this chapter is to describe a complete derivation of the FFT-based method for KDE bandwidth selection. First, some important limitations of the existing solution are emphasized and crucial modifications are proposed. Then, the next step shows how the FFT-based method can be used for very fast KDE computations, as well as for bandwidth selection. Following that, we present a detailed derivation of the *least square cross validation* bandwidth selector using the FFT-based method. We also discuss the use of this method for plug-in and *smoothed cross validation* bandwidth selectors. The final part is devoted to an overview of extended computer simulations confirming high performance and accuracy levels of the FFT-based method for KDE and bandwidth selection.

FPGA-based implementation of a bandwidth selection algorithm. Chapter 6 discusses author's own research related to fast computation using the univariate plug-in algorithm, a type of a bandwidth selection algorithm. In contrast to the results presented in Chap. 5, this chapter describes a hardware-based method, which relies on utilizing the so-called *field-programmable gate arrays* (FPGA). FPGA devices are not often used for purposes of implementing purely numerical algorithms. The proposed implementation can be seen as a preliminary assessment of practical usability of such FPGA-based applications. We describe the notion of a *high level synthesis* (HLS) approach and then, move on to rewrite the plug-in algorithm in a way ready for a direct FPGA/HLS implementation. This is followed by a description of implementation preliminaries with certain concepts then being described in more detail. The final part of this chapter describes the results confirming the practical usability of FPGA chips for fast implementations of complex numerical algorithms.

Selected application related to kernel density estimation. Chapter 7 presents a number of applications related to KDE. The first one is *discriminant analysis*, a well-known data exploration method. The second is *cluster analysis* (this is also a well-researched field), where the so-called *mean-shift* algorithm is used. We illustrate the two areas with some simple numerical examples confirming the practical usability of these KDE-based variants of the algorithms used. Next, the *nonparametric kernel regression* is presented. It can be viewed as an interesting alternative to the classical parametric regression techniques. The fourth application is *multidimensional statistical process control*. Here, a kernel-based approach is a worth considering option if the underlying d-variate process is not multivariate normal. The final part is devoted to presenting a complete framework for the so-called *gating* procedure widely used in analyzing *flow cytometry* datasets. The framework is based on a smart adaptation of the so-called *feature significance* technique. To show that it can be used in practical terms, we provide a numerical example based on a real flow cytometry dataset. The described results show that the proposed method can be considered an alternative to classical gating methods.

1.3 Topics Not Covered in This Book

It goes without saying that this book does not cover every topic related to KDE. We round off this chapter with a short list of the main problems that are not described in this book (the details can be found in the references provided).

Kernel methods for data streams. A variety of real-live applications rely on analyzing outputs that are generated as (often infinite) data streams. Unfortunately, when the data comes as a stream, traditional density estimation methods cannot be used. The main problem which appears here is that a large volume of data 'flows' continuously and every individual datum is available for analysis only for a short period. Moreover, usually it is either impossible or impractical to store (accumulate) all the coming data permanently. Consequently, dedicated methods are required to generate density estimations over data streams [16, 90, 200].

Kernel regression smoothing. Regression is one of the most often used statistical techniques with linear modeling, forming a subclass of this technique, is probably the most widely used tool. However, there are many situations, when such modeling cannot be applied, mainly as a result of strong nonlinearity in the data. Kernel regression smoothing is a problem that is closely related to KDE. Here, the goal is to estimate the conditional expectation of a random variable ($\mathrm{E}(Y|X) = m(X)$, with m being an unknown function). The objective is to find a non-linear relation between a pair of random variables X and Y. The Nadaraya-Watson estimator is a very common choice [81, 124, 193]. In Sect. 7.4, only a small example of the use of this technique is described.

Kernel methods for categorical and binary data. Smoothing techniques are usually applied to continuous data, hence a smooth density estimate $\hat{f}$ is typically generated in consequence. The traditional nonparametric kernel methods work on the assumption that the underlying data is continuous. However, practical datasets can contain mix of categorical (unordered and ordered variables) and continuous data types, as well as binary data (numeric values using two different symbols: typically 0 and 1). Kernel methods can also be used for such data [88, 137, 169, 170, 180].

Nonparametric estimation of various statistical characteristics. Kernel methods can be used not only to estimate the probability density functions. It can also be applied to such characteristics as the cumulative distribution function (briefly presented in Sect. 3.7), quantiles, receiver operating characteristic curves (ROC curves), hazard functions, survival functions and others, see [48, 94, 107, 109, 110].

Chapter 2
Nonparametric Density Estimation

2.1 Introduction

This chapter describes the background material related to the nonparametric density estimation. Techniques such as histograms (together with its extension, known as ASH, see Sect. 2.3), Parzen windows and k-nearest neighbors are at the core of the applications of nonparametric density estimation. For that reason, we decided to include a chapter describing these for the sake of completeness and to allow less experienced readers develop their intuitions in terms of the nonparametric estimation.

Most of the material is presented taking into account only the univariate case; extending the results to cover more than one variable, however, is often a straightforward task.

The chapter is organized as follows: Sect. 2.2 presents a short overview of the fundamental concepts related to histograms. Section 2.3 is devoted to a description of a smart extension of certain well-known histograms aimed at avoiding some of their drawbacks. Section 2.4 presents basic concepts related to the nonparametric density estimation. Section 2.5 is devoted to the Parzen windows, while Sect. 2.6 to the k-nearest neighbors approach.

2.2 Density Estimation and Histograms

The well-known histogram is the simplest form of a nonparametric density estimation. The sample space is divided into disjoint categories, or *bins*, (note that the number of bins is expressed by a natural number) and the density is approximated by counting how many data points fall into each bin. Let B_l be the l-th bin and h be the width of the bins (all the bins have equal widths) and let $\#\{X_i \in B_l\}$ denote the total number of data points from $X_1, X_2, \ldots, X_n$ that fall into the corresponding bin B_l of width h. Then, the PDF estimator is

A. Gramacki, *Nonparametric Kernel Density Estimation and Its Computational Aspects*, Studies in Big Data 37,
https://doi.org/10.1007/978-3-319-71688-6_2

$$\hat{f}(x) = \frac{\#\{X_i \in B_l\}}{nh} = \frac{k}{nh}, \tag{2.1}$$

for every $x \in B_l$, with l being a natural number. The above uses a common convention, namely that $f(x)$ and $\hat{f}(x)$ represent the true (usually unknown) density and an estimator of the true density, respectively. Figure 2.1 shows a sample histogram generated for a toy univariate dataset of seven data points:

$$X_1 = 3.5,\ X_2 = 4.2,\ X_3 = 4.5,\ X_4 = 5.8,\ X_5 = 6.2,\ X_6 = 6.5,\ X_7 = 6.8. \tag{2.2}$$

In this example, $h = 1$ and the starting position of the first bin (also known as the *bin origin*) is $x_0 = 0$. Note also that bins B_1 and B_6 are empty.

As it can be easily seen, the histogram requires two parameters to be defined: the bin width h and the bin origin x_0. While the histogram is a very simple form of the nonparametric density estimator, there are some serious drawbacks that are already noticeable. First, the final shape of the density estimate strongly depends on the starting position of the first bin. Second, the natural feature of the histogram

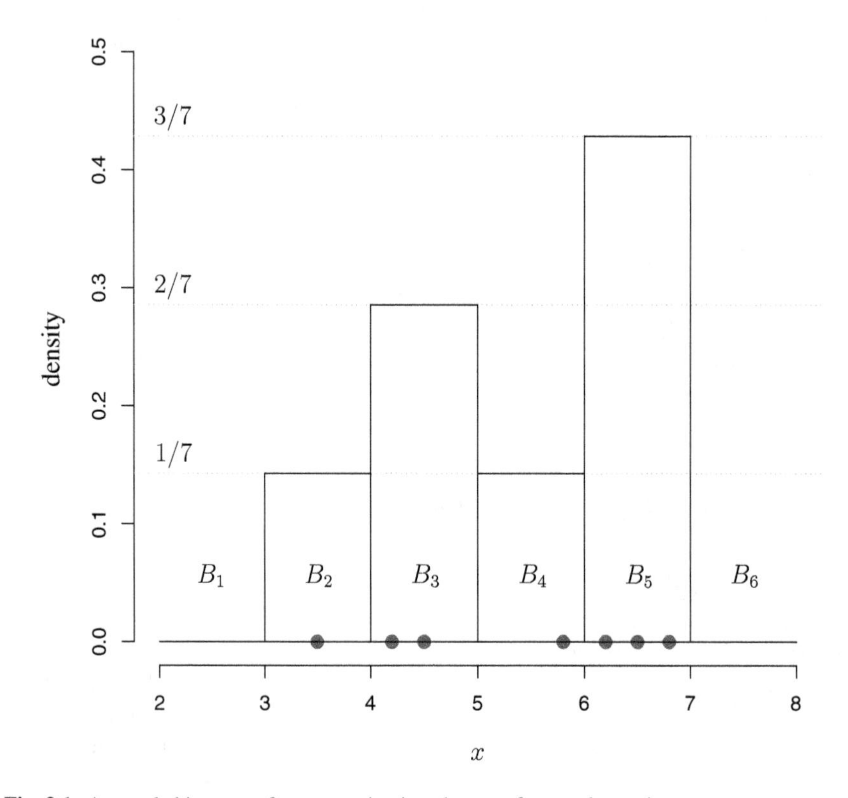

Fig. 2.1 A sample histogram for a toy univariate dataset of seven data points

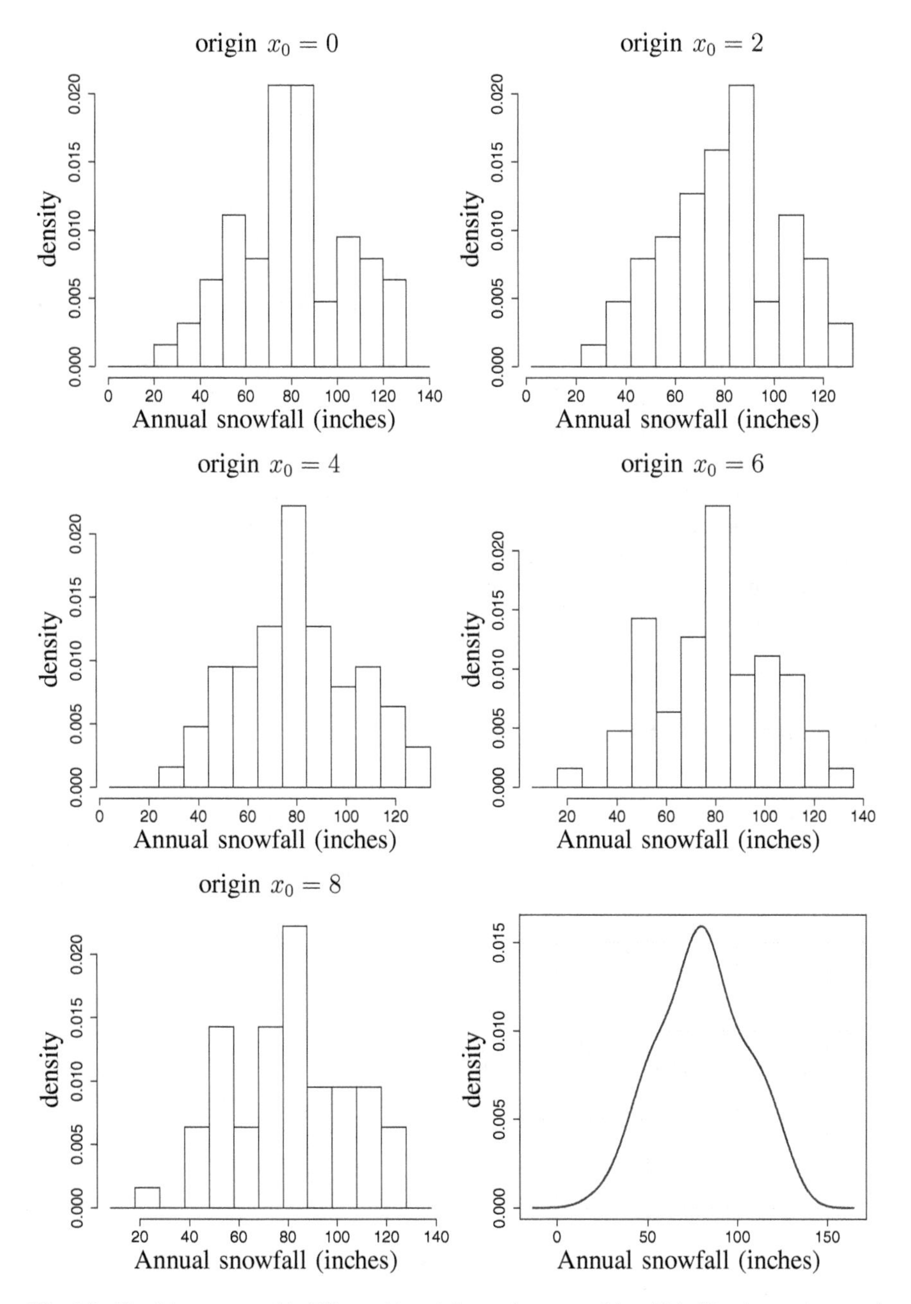

Fig. 2.2 Five histograms with different bin origins and constant bin width. The dataset is annual snowfall accumulations in Buffalo, NY from 1910 to 1973 (in inches)

is the presence of discontinuities of density. These are not, however, related to the underlying density and, instead, are only an artifact of the chosen bin locations. Third drawback is the so-called curse of dimensionality (see Sect. 3.9), which constitutes a much more serious problem, since the number of bins grows exponentially with the number of dimensions. In higher dimensions one would require a very large number of examples or else most of the bins would be empty (incidentally, the curse of dimensionality phenomena is a common problem for all the nonparametric techniques for density estimation). All these drawbacks make the histogram unsuitable for most practical applications except for rapid visualization of results in one or two dimensions (less often in three dimensions).

Figure 2.2 illustrates the phenomenon of the strong dependence of the histogram shape on the starting position of the first bin (here the *buffalo* dataset from the *gss* *R* package was used). To showcase this, the bin width remains constant ($h = 10$), while different origins are used ($x_0 = 0, 2, 4, 6, 8$). It is not obvious which histogram should be considered as the best one. All five histograms suggest a mode around the value of 80, but in some cases the existence of two or even three modes is not excluded. In the lower right corner the 'true' density is depicted, generated with a smooth nonparametric kernel density estimator. This shows that, in fact, only one mode is present in the input data.

2.3 Smoothing Histograms

As demonstrated in the previous section, the bin origin (sometimes called a *nuisance parameter*) has a significant influence on the final histogram shape. A smart extension of the classical histogram was presented in [156, 159], referred to as the *averaged shifted histogram* (ASH). ASH enjoys several advantages compared with the classical histogram: better visual interpretation, better approximation, and nearly the same computational efficiency as classical histograms. ASH provides a bridge between the classical histogram and advanced kernel-based methods presented in Chap. 3.

ASH algorithm avoids the pitfall of choosing an arbitrary value for the bin origin x_0. It is a nonparametric density estimator that averages several classical histograms with different origins. A collection of m classical histograms, each with the bin width h, but with slightly different (or shifted) origin is constructed. Then, the average of these histograms is calculated and the ASH estimate at x is then defined as

$$\hat{f}_{ASH}(x) = \frac{1}{m} \sum_{i=1}^{m} \hat{f}_i(x). \tag{2.3}$$

Figure 2.3 shows the example ASH density estimates of a sample dataset generated from a mixture of three Gaussians given by

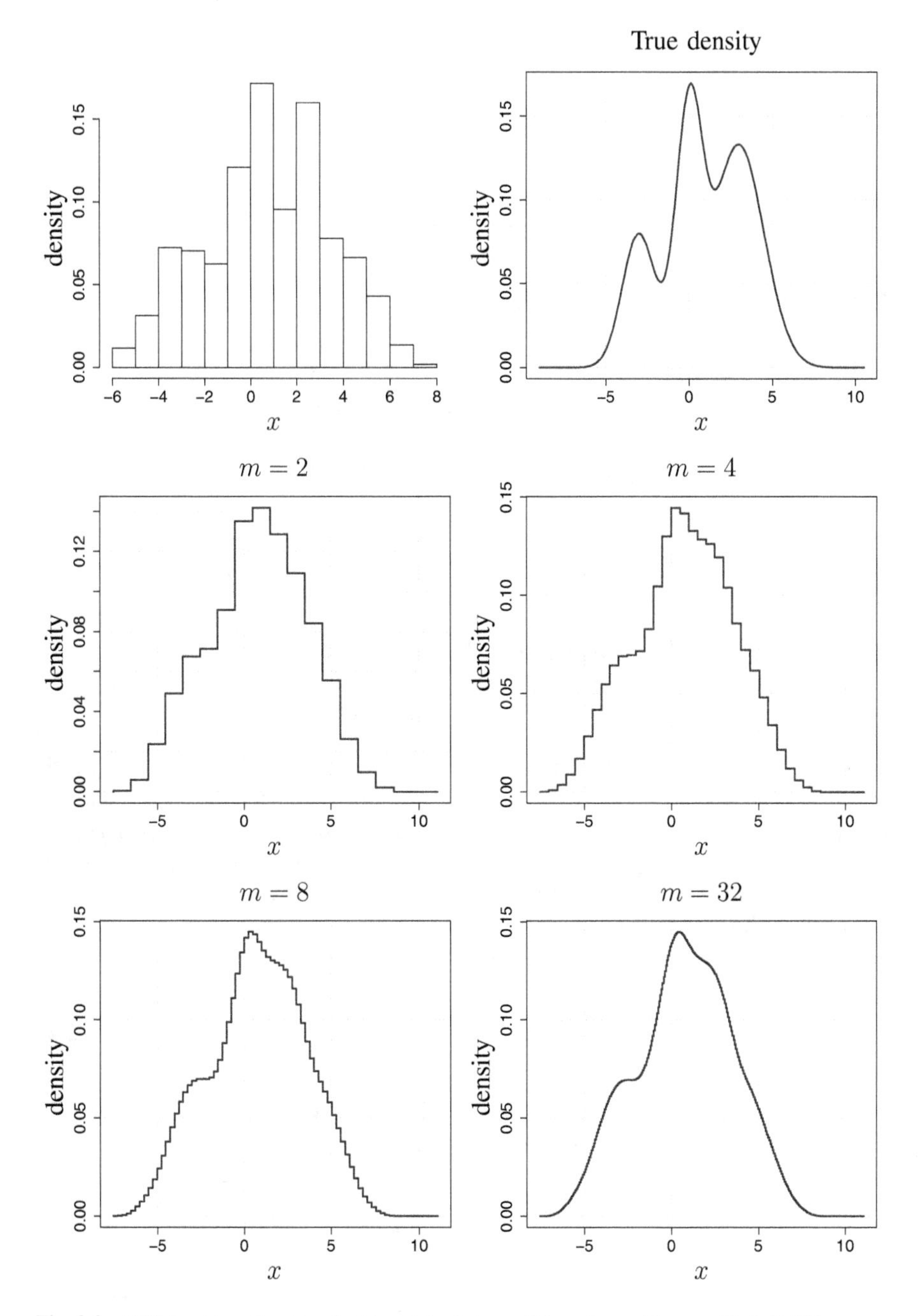

Fig. 2.3 ASH density estimates of a trimodal mixture of Gaussians with $m = \{2, 4, 8, 32\}$

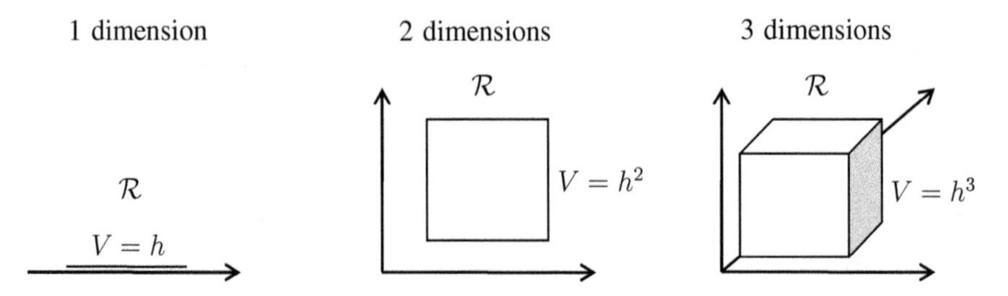

Fig. 2.4 Hypercubes in one, two and three dimensions

$$f(x) = \frac{2}{10}\mathcal{N}(x;\ -3,\ 1) + \frac{3}{10}\mathcal{N}(x;\ 0,\ 0.8) + \frac{5}{10}\mathcal{N}(x;\ 3,\ 1.5)\,. \tag{2.4}$$

It can be easily observed that already for m at the level of several dozen, the resulting density is smooth and accurate enough with the trimodal nature of the data being evident.

2.4 A General Formulation of Nonparametric Density Estimation

This section aims at developing intuitions related to the nonparametric density estimation. Assume that n samples $X_1, X_2, \ldots, X_n$ are drawn independently and are identically distributed (*iid*) from a (usually unknown) density function p. The goal is to estimate this density at an arbitrarily chosen point x based on these n samples.

Now, let $\mathcal{R}$ be a region around x. A region $\mathcal{R}$ is considered to be a d-dimensional hypercube with side length h and volume $V = h^d$, as depicted in Fig. 2.4.

The probability that a training sample will fall in a region $\mathcal{R}$ is

$$P = \Pr[x \in \mathcal{R}] = \int_{\mathcal{R}} p(x)dx. \tag{2.5}$$

It is well known that the probability that k of these n samples fall in $\mathcal{R}$ is given by the binomial distribution

$$\Pr(k;\, n,\, p) = \Pr(X = k) = \binom{n}{k} P^k(1 - P)^{n-k}. \tag{2.6}$$

It is also known, given the properties of the binomial distribution, that the mean and the variance of the ratio k/n are described by the following equations

$$\mathrm{E}\left(\frac{k}{n}\right) = P,$$
$$\mathrm{Var}\left(\frac{k}{n}\right) = \frac{P(1-P)}{n}. \tag{2.7}$$

Now, it should be clear that the variance approaches zero as $n \to \infty$. So, it can be expected that the mean fraction of points that fall within $\mathcal{R}$, that is

$$P \cong \frac{k}{n}, \tag{2.8}$$

would be a good estimate of the probability P. On the other hand, if it is assumed that $\mathcal{R}$ is so small that the density $p(x)$ does not vary too much within it, then the integral in (2.5) can be approximated by

$$P = \int_{\mathcal{R}} p(x)dx \cong p(x)V, \tag{2.9}$$

where V is the volume of the region $\mathcal{R}$ (for example the volume V being the width of the gray area in Fig. 1.1, is obviously too wide to satisfy the conditions of (2.9)). Finally, by merging (2.8) and (2.9) we obtain that

$$p(x) \cong \frac{k}{nV}. \tag{2.10}$$

Equation (2.10) converges in probability to the true density according to the following

$$f(x) = \lim_{n\to\infty} p(x) = \lim_{n\to\infty} \frac{k}{nV}. \tag{2.11}$$

The following conditions are required for convergence

$$\lim_{n\to\infty} V_n = 0,$$
$$\lim_{n\to\infty} k_n = \infty,$$
$$\lim_{n\to\infty} \frac{k_n}{n} = 0. \tag{2.12}$$

The estimate becomes more accurate as the number of samples n increases and the volume V shrinks. In practical applications, the value of n is always fixed (the input datasets are unchangeable). So, to improve the estimate of $f(x)$ one could set V to be sufficiently small. In that case however, the region $\mathcal{R}$ can become so small that it would enclose no data. This means that in practice, a kind of compromise in terms of finding a proper value of V is needed. It should be large enough to include a sufficient number of samples within $\mathcal{R}$ and, at the same time, be small enough to satisfy the

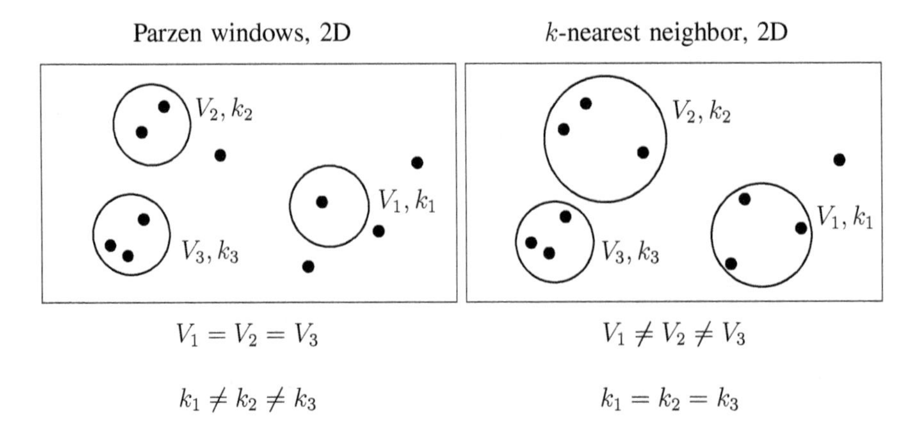

Fig. 2.5 A visualization of the idea of the Parzen windows and the k-nearest neighbors techniques

assumption of (2.9). This is a typical *trade-off dilemma* encountered when dealing with different knowledge domains.

Equation (2.10) can be regarded as a starting point for every nonparametric density estimation analysis. It is sometimes referred to as the *naive density estimator*. Obviously, if $V = h$, the Eq. (2.10) is the same as the one in the definition of the histogram (2.1).

In practical situations related to the use of density estimators two basic approaches can be distinguished:

- *Parzen windows* [130]: choosing a fixed value of the volume V and determining the corresponding k directly from the data,
- *k-nearest neighbors* (in [131] it is reported that the first mention of the k-nearest neighbor technique was given in [62]): choosing a fixed value of k and determining the corresponding volume V directly from the data.

The general idea of the above is visualized in Fig. 2.5. It can be proved that both approaches converge in probability to the true density $f(x)$ as $n \to \infty$, assuming that volume V shrinks with n, and k grows with n, appropriately. The parameter k is usually called the *smoothing parameter* or the *bandwidth*. In general, the estimation of its optimal value is not trivial.

2.5 Parzen Windows

This section provides a brief introduction to the first approach (Parzen windows) for constructing nonparametric density estimators. Suppose that the region $\mathcal{R}$ is a hypercube of side length h and it encloses the k samples. Then, it should be obvious that its volume V is given by $V = h^d$, where d is the dimensionality of the problem.

To estimate the unknown density $f(x)$ at a point x, simply center the region $\mathcal{R}$ at this point x and count the number of samples k in $\mathcal{R}$ and then substitute this value tos (2.10). The above can be expressed analytically by defining a *kernel function* (or a *window function*)

$$K(u) = \begin{cases} 1 & |u_i| \leq 1/2 \quad \forall i = 1, \ldots, d \\ 0 & \text{otherwise,} \end{cases} \tag{2.13}$$

and can be visualized (in one and two dimensions) as shown in Fig. 2.6. This kernel corresponds to a unit hypercube centered at the origin and is known as the *Parzen window*.

Now, to count the total number of points k from the input dataset $X_1, X_2, \ldots, X_n$ that are inside the hypercube with side length h centered at x the following expression can be used

$$k = \sum_{i=1}^{n} K\left(\frac{x - X_i}{h}\right). \tag{2.14}$$

Taking into account (2.10), the desired analytical expression for the estimate of the true density $f(x)$ can be formulated as follows

$$\hat{f}(x) = \frac{1}{nh^d} \sum_{i=1}^{n} K\left(\frac{x - X_i}{h}\right). \tag{2.15}$$

It is evident that the Parzen windows method is very similar to the histogram with the only exception that the bin origins are determined by the input data itself. Equation (2.13) can obviously be rewritten as

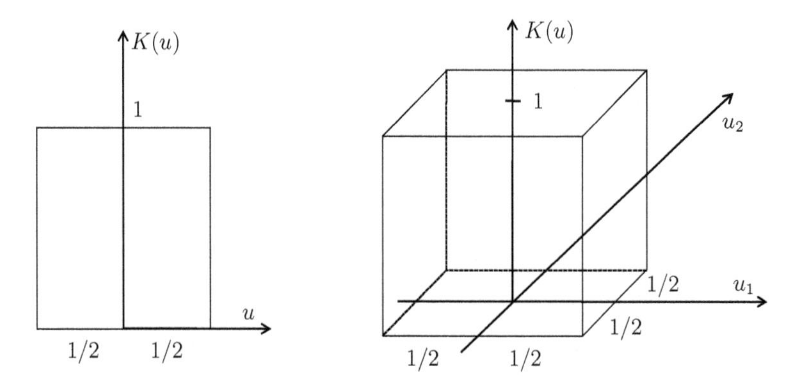

Fig. 2.6 Window functions in one and two dimensions

$$K\left(\frac{x - X_i}{h}\right) = \begin{cases} 1 & |x - X_i| \leq h/2 \quad \forall i = 1, \ldots, d \\ 0 & \text{otherwise.} \end{cases} \tag{2.16}$$

Figure 2.7 demonstrates the construction of the Parzen windows estimator. It is a simple observation that the Parzen windows density estimator can be considered as a sum of boxes (for general multivariate case—hyperboxes) centered at the observations. Figure 2.8 shows several Parzen windows estimates of a mixture of two Gaussians given by

$$f(x) = \frac{1}{2}\mathcal{N}(4, 1) + \frac{1}{2}\mathcal{N}(7, 0.5). \tag{2.17}$$

The resulted densities are very jagged, with many discontinuities. This example uses the *normal scale* selector introduced in [168] described by the following equation

$$h_{NS} = \left(\frac{4}{n(d+2)}\right)^{1/(d+4)} \sigma, \tag{2.18}$$

where σ is the standard deviation of the input data.

Coming to the end of this section, we should also point out that the Parzen windows methodology has a few serious disadvantages. First, the density estimates have discontinuities and are not smooth enough. Second, all the data points are weighted

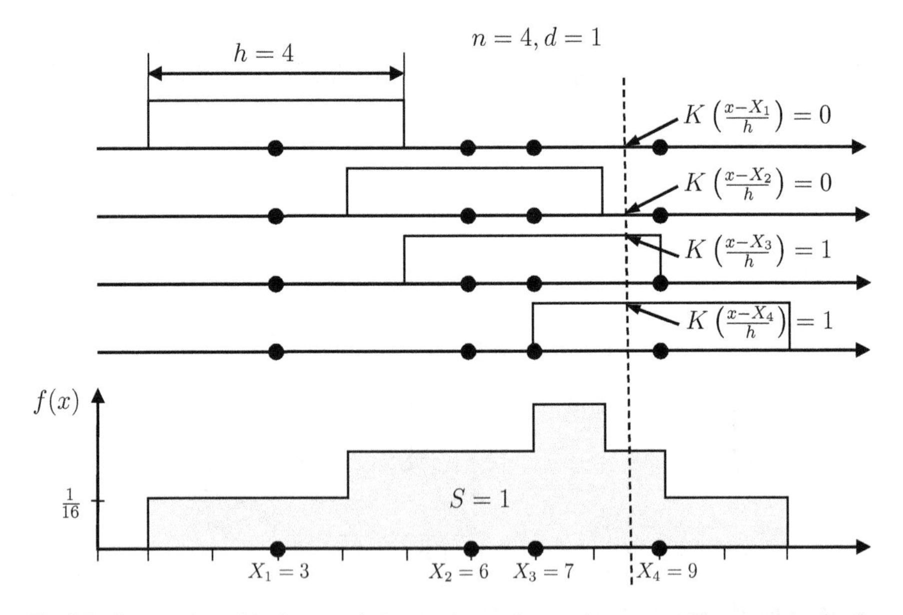

Fig. 2.7 Construction of the Parzen windows estimator for $n = 4$ as a sum of boxes centered at the observations

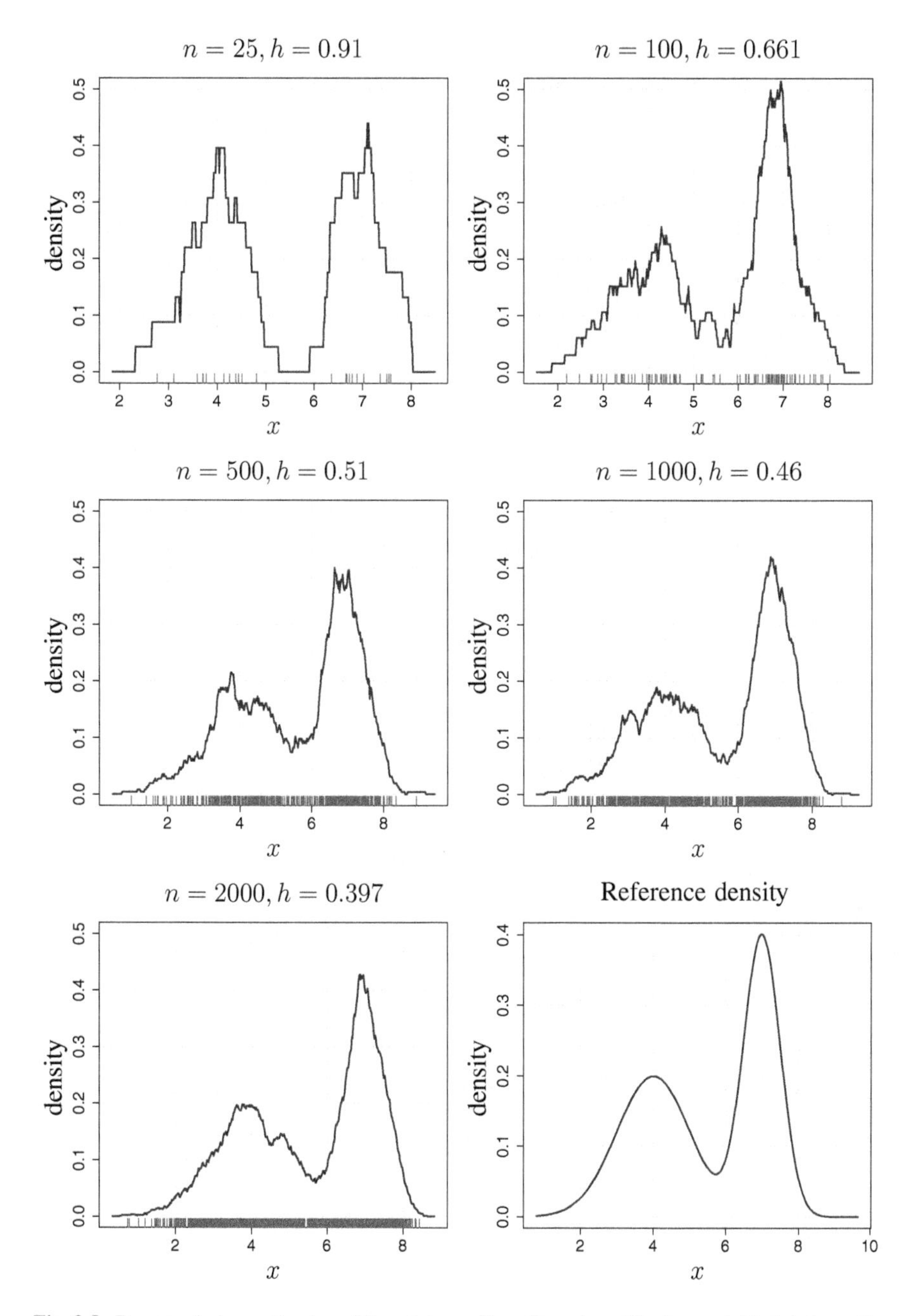

Fig. 2.8 Parzen window estimates of the mixture of two Gaussians. The lower right plot shows the true density curve

equally, regardless of their distance to the estimated point x (kernel function is ‘flat’). Third, one might need a larger number of samples in order to obtain accurate estimates of the searched density.

2.6 k-nearest Neighbors

This section briefly describes the second approach (k-nearest neighbors) for constructing nonparametric density estimators. In the k-nearest neighbors (KNN) method, a fixed value of k is chosen and the corresponding volume V is determined directly from the data. The density estimate now is

$$p(x) = \frac{k}{nV} = \frac{k}{nc_d R_k^d(x)}, \tag{2.19}$$

where $R_k^d(x)$ is the distance between the estimation point x and its k-th closest neighbor and c_d is the volume of the unit sphere in d dimensions given by

$$c_d = \frac{\pi^{d/2}}{(d/2)!} = \frac{\pi^{d/2}}{\Gamma(d/2+1)}. \tag{2.20}$$

This is depicted in Fig. 2.9 (for $d = 2$ and $k = 4$).

In general, the estimates generated by KNN are not considered to be adequate. The main drawbacks are as follows:

- these estimates are very prone to local noises,
- these estimates are far from zero (have very heavy tails), even for large regions where no data samples are present,

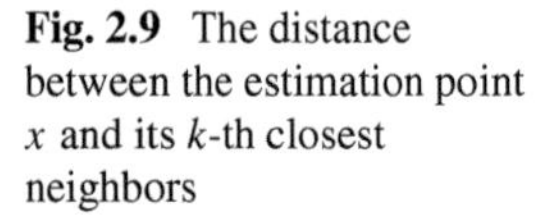
Fig. 2.9 The distance between the estimation point x and its k-th closest neighbors

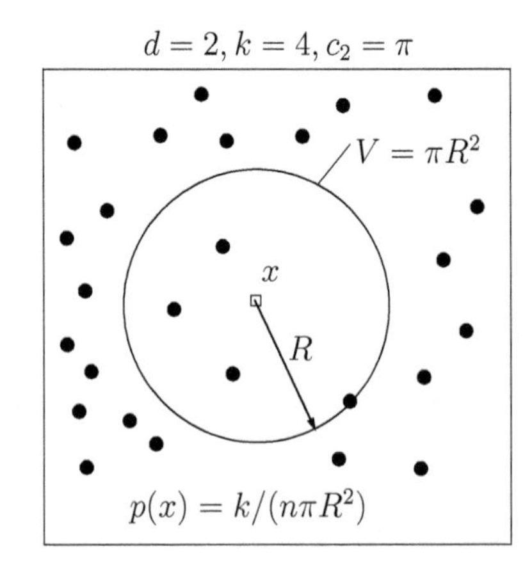

- these estimates have many discontinuities (look very spiky, are not differentiable),
- the condition (1.3) does not hold (in fact no 'legal' densities are obtained).

The last point can be demonstrated using a toy univariate example where only one data point is present, i.e. $n = 1$. k is estimated using a popular rule-of-thumb saying that $k = \sqrt{n}$ (and rounded to the nearest integer). Here one obtains (see (2.19))

$$p(x) = \frac{k}{nV} = \frac{1}{2|x - X_i|}. \tag{2.21}$$

In this example, it is obvious that

$$\int_{-\infty}^{+\infty} \frac{1}{2|x - X_i|} = \infty \neq 1. \tag{2.22}$$

Figure 2.10 shows a very simple toy example where four datasets contain only 1, 3, 4 and 5 points respectively (marked as small black-filled circles). It goes without saying that such dataset sizes are of no practical importance in terms of any plausible nonparametric density estimation. Nevertheless, this simple example succeeds in demonstrating the above-mentioned drawbacks of the KNN estimator. The values of the integral (1.3) are 7.51, 1.27, 1.33 and 1.56 respectively (the integral limits were narrowed down to the values roughly the same as the x-label limits shown in the plots, so the value of the integral in the upper left plot is smaller than ∞).

Figure 2.11 shows several KNN estimates of a mixture of two Gaussians given by (2.17). The resulting densities (which are in fact not true densities, see the considerations above) are very jagged, with too heavy tails and many discontinuities. The values of the integral (1.3) are 0.93, 1.33, 1.23, 1.24 and 1.20 for sample datasets with $n = \{16, 32, 64, 256, 512\}$, respectively (the integral limits are narrowed down to the values roughly the same as the x-label limits shown in the plots). This example uses a more sophisticated k selector, introduced in [50], given by the following equation

$$\begin{aligned} k_{opt} &= v_0 \left[\frac{4}{d+2}\right]^{d/(d+4)} n^{4/(d+4)}, \\ v_0 &= \pi^{d/2}\Gamma((d+2)/d), \end{aligned} \tag{2.23}$$

where v_0 is the hyper-volume of the unit d-ball. Figure 2.12 shows several KNN estimates of a mixture of two Gaussians given by (2.17) for two different n values and four different k values. Note that for $n = 100$ the optimal k is 37 and for $n = 250$ the optimal k is 78 (as calculated using (2.23)).

A question naturally suggests itself at this point whether one should give up the KNN approach completely. The answer is 'no'. Typically, instead of approximating unknown densities $f(x)$, the KNN method is often used for the classification task and it is a simple approximation of the (optimal) Bayes classifier [44].

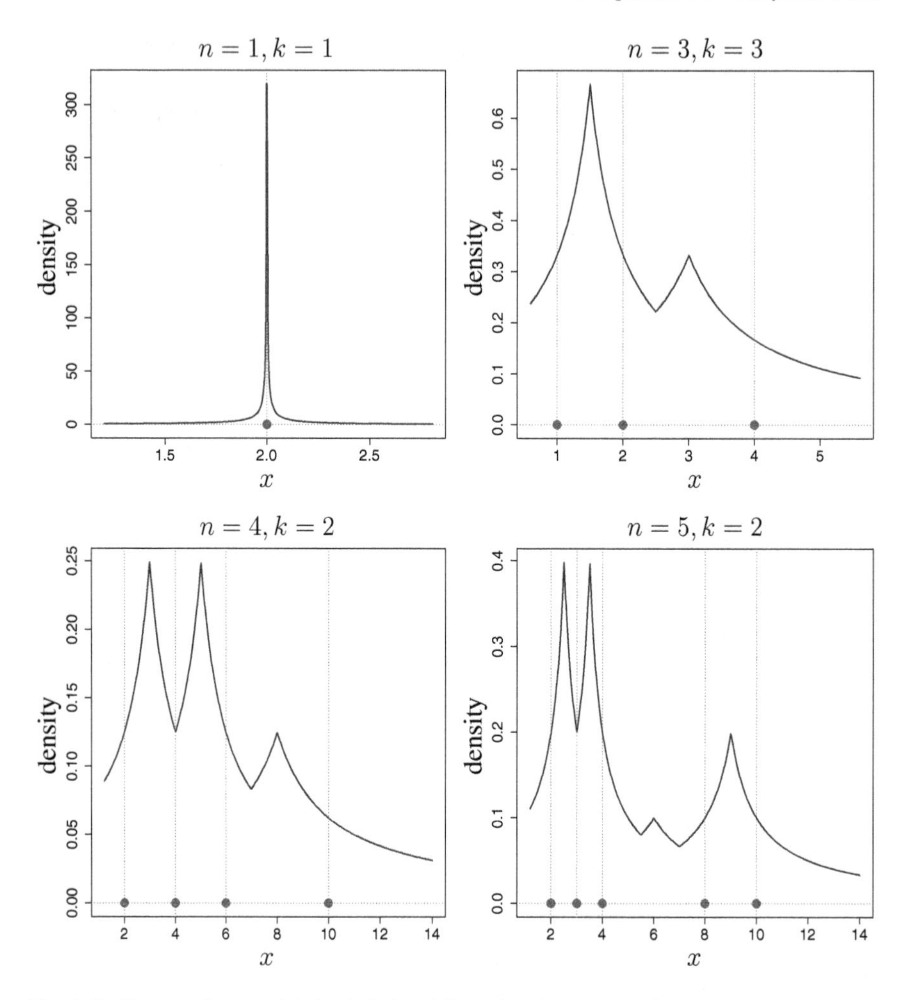

Fig. 2.10 Four toy datasets (of size 1, 3, 4 and 5) used to demonstrate the main idea behind KNN estimators

The idea here is to classify a new sample using a form of *majority voting*. Figure 2.13 is an example of that approach. The training samples, marked by filled circles, belong to the first class and the samples marked by open circles belong to the second class. Here, a 5-NN classifier ($k = 5$ nearest neighbors) is used: that is a circle is added to enclose the five nearest neighbors of the new sample marked by the star sign. The task is to decide to which class this new instance should be assigned. The 5-NN classifier would classify the star to the first class (filled circles), since the majority of the objects (three out of five) located inside the circled region are filled circles.

To put the above in more general terms, assume that one has a dataset with n examples, with n_i of them coming from the class c_i so that $\sum_i n_i = n$. The goal is

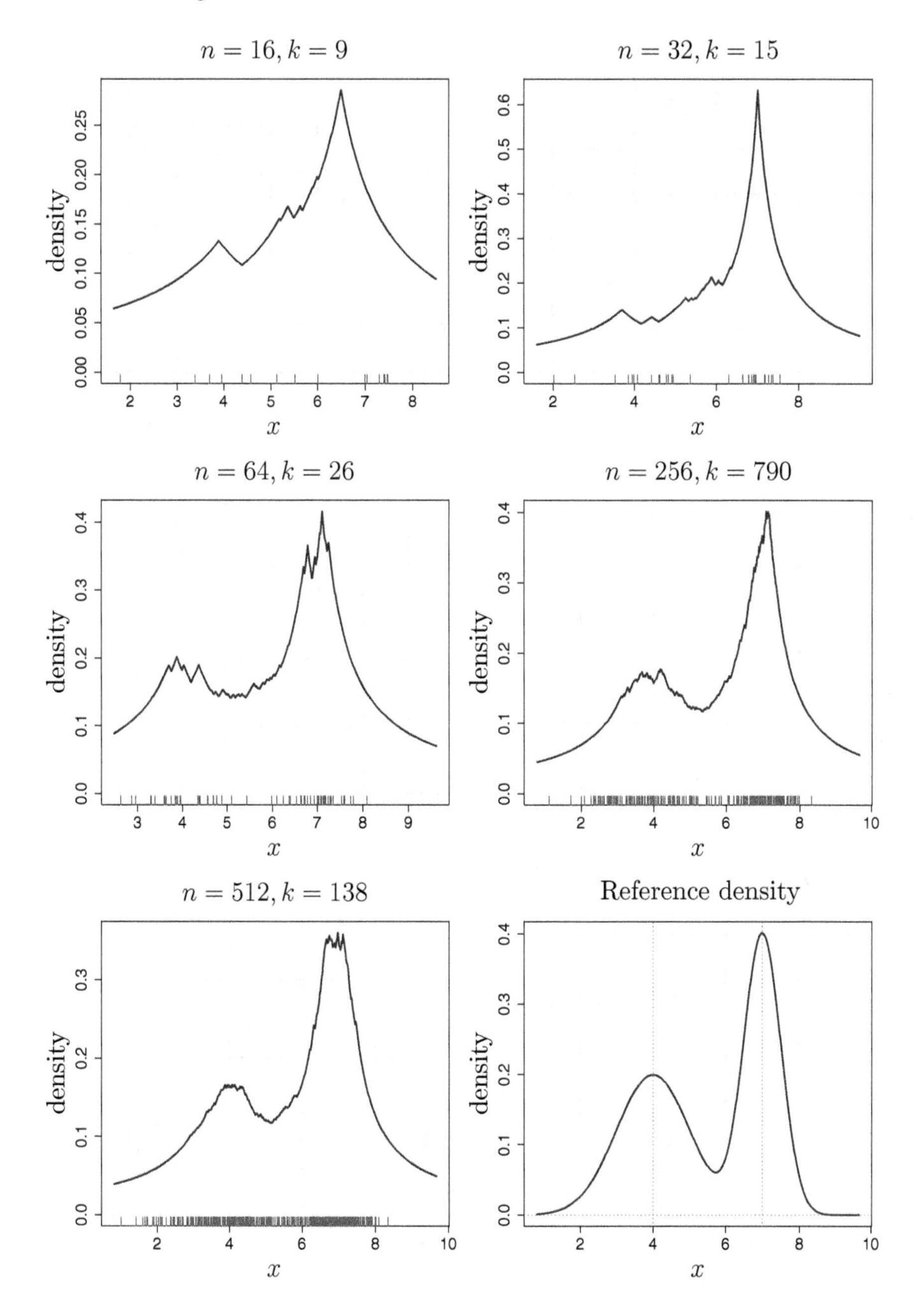

Fig. 2.11 KNN estimates of a mixture of two Gaussians. The lower right plot shows the true density curve

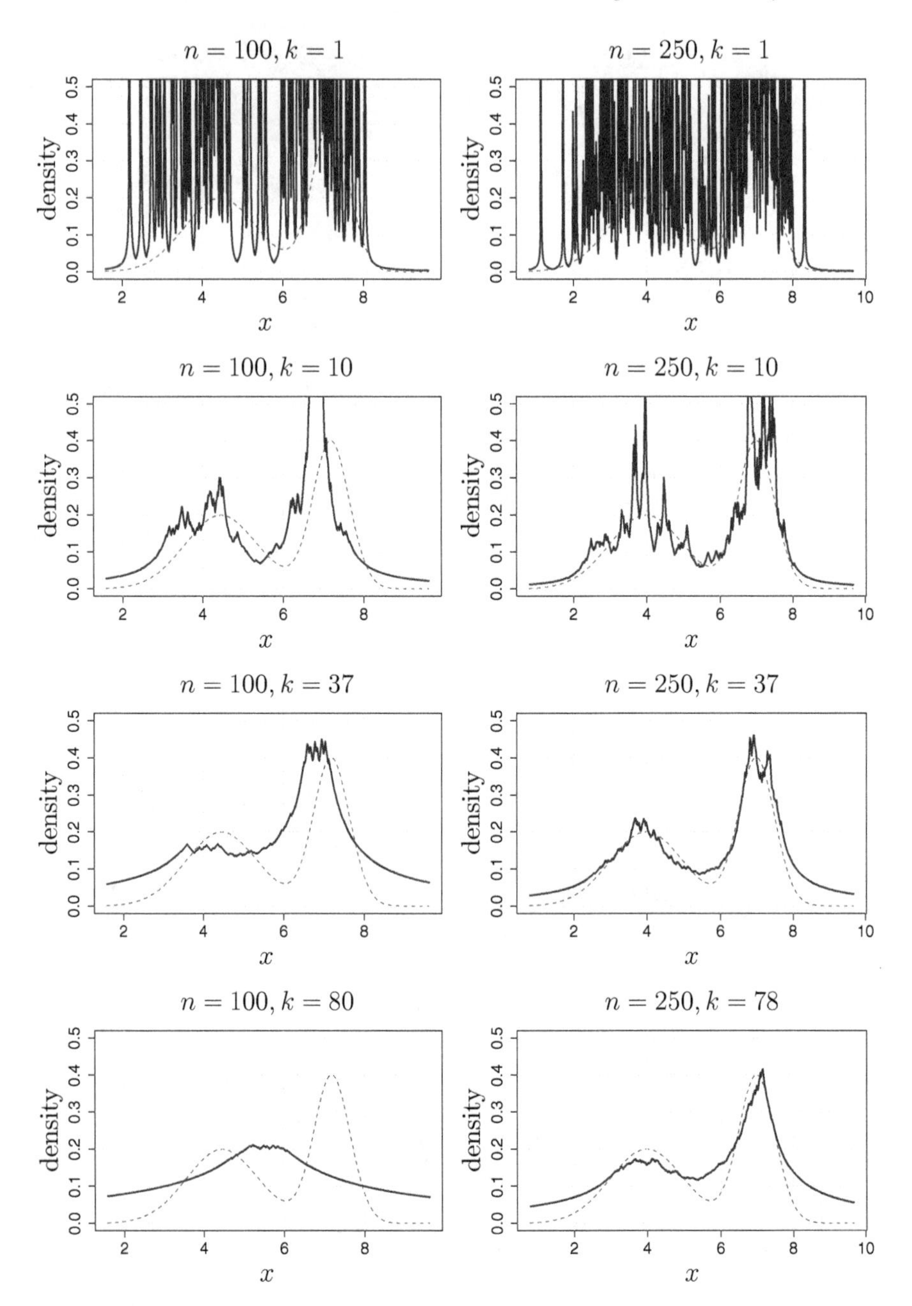

Fig. 2.12 KNN estimates of a mixture of two Gaussians. Different k and n are used. For $n = 100$ the optimal k is 37 and for $n = 250$ the optimal k is 78. The true density curve is plotted with the dashed line

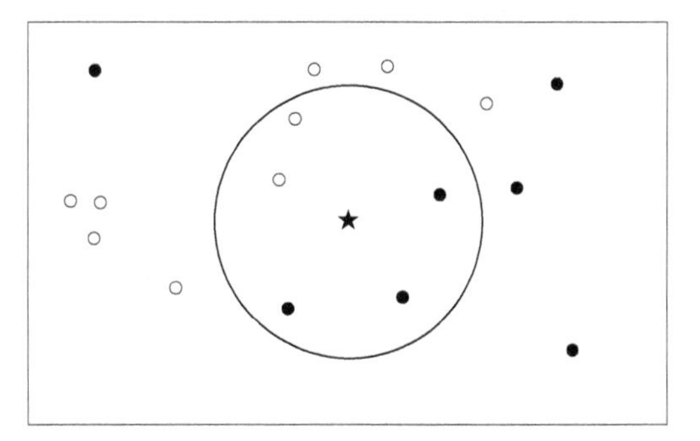

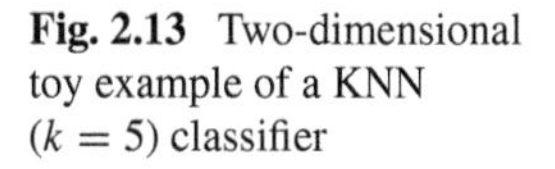

Fig. 2.13 Two-dimensional toy example of a KNN ($k = 5$) classifier

to classify an unknown sample x. Recall that the unconditional density (now labeled $p(x)$) is estimated by $p(x) = k/(nV)$. Place a hyper-sphere with volume V around x enclosing k samples. Let k_i samples out of k be labeled c_i, then we have that

$$p(x|c_i) \cong \frac{k_i}{n_i V}, \tag{2.24}$$

and the prior probabilities (also known as class priors) are approximated by

$$p(c_i) \cong \frac{n_i}{n}. \tag{2.25}$$

Now, putting everything together, and using Bayes' theorem one obtains the posterior probability of the class membership using

$$p(c_i|x) = \frac{(x|c_i)p(c_i)}{p(x)} = \frac{\frac{k_i}{n_i V}\frac{n_i}{n}}{\frac{k}{nV}} = \frac{k_i}{k}. \tag{2.26}$$

The estimate of the posterior probability is provided simply by the fraction of samples that belong to the class c_i. This is a very simple and intuitive estimator. In other words, given an unlabeled example x, find its k closest neighbors among the example labeled points. Then, assign x to the most frequent class among these neighbors (as demonstrated in Fig. 2.13). The KNN classifier only requires the following:

- an integer k (determined using for example (2.23)),
- a set of labeled examples (as training data, used in the verification process),
- a metric to measure the distance between x and potential clusters. Usually, the classical Euclidean distance is used.

Finally, in Fig. 2.14 a simple example based on an artificial dataset is presented (a Gaussian mixture model).

The training samples belong to two classes, denoted by open and filled circles, respectively. Based on this training dataset, four KNN classifiers have been built (for

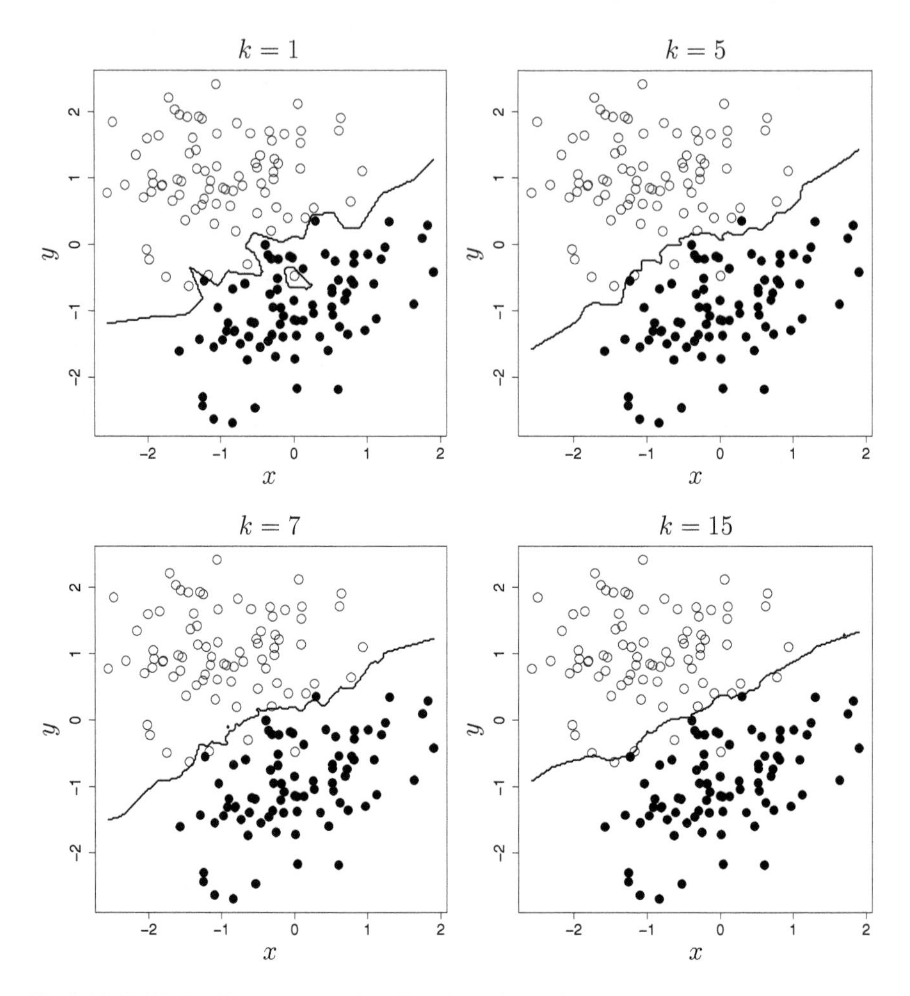

Fig. 2.14 KNN classifiers on a two-class Gaussian mixture data

$k = \{1, 5, 7, 15\}$) with decision boundaries represented by solid lines. It is easily seen how k affects the boundaries. The problem of selecting the optimal k is a critical one in terms of obtaining the final results. A small value of k means that noise can have a higher influence on the result (risk of overfitting). On the other hand, a large value of k makes it computationally expensive and the KNN classifier may misclassify the test sample because its list of nearest neighbors can include many samples from other classes (risk of oversimplifying). Moreover, more numerous classes can have a dominant impact on assigning x to these classes. A popular rule of thumb in choosing the value of k is to set $k = \sqrt{n}$, where n is the number of samples in the training dataset. Another popular method is based on certain cross-validation techniques. This approach however, is often unsuccessful as reported in [65].

Chapter 3
Kernel Density Estimation

3.1 Introduction

Chapter 2 presented basic notions related to the nonparametric density estimation. We stated that the k-nearest neighbors method has a very limited scope of practical applications (in the context of KDE), mainly due to its very poor performance. In turn, the Parzen windows method presented in Sect. 2.5 gives slightly better performance but it also produces discontinuities (stair-like curves) that are quite annoying in practice. Fortunately, this drawback can be easily eliminated by using a proper *smooth kernel function*, other than (2.13). A popular choice for K is the *normal (Gaussian) density function* that is radially symmetric and unimodal. This kernel is, by all accounts, the most often used one in terms of practical applications, however, there are other kernels subject to research, as evidenced in the relevant literature.

The books [13, 158, 168] are classical positions describing the notions related to kernel density, one also should note a number of more recent publications e.g. [84, 94, 107] describing the topic. Many authors present different aspects of KDE and its practical applications, see for example [26, 41, 51, 107, 108, 125, 152, 171] and many others.

The chapter is organized as follows: Sect. 3.2 briefly presents the most popular univariate kernels used in KDE. Section 3.3 is a brief description of the relevant definitions used in the context of KDE, both for uni- and multivariate cases. Section 3.4 is devoted to performance criteria that constitute a background for a deeper KDE analysis, including a very important problem of bandwidth selection. Sections 3.5, 3.6, 3.7 and 3.8 contain information on a few selected extensions of the basic KDE. Section 3.9 presents a problem known in literature as the curse of dimensionality, an inherent feature of most of the nonparametric methods. Some computational considerations related to KDE are presented in Sect. 3.10.

A. Gramacki, *Nonparametric Kernel Density Estimation and Its Computational Aspects*, Studies in Big Data 37,
https://doi.org/10.1007/978-3-319-71688-6_3

3.2 Univariate Kernels

The univariate normal kernel is described by the following well-known formula

$$K(u) = (2\pi)^{-1/2}\exp\left(-\frac{1}{2}u^2\right). \tag{3.1}$$

The kernel to be used in KDE must satisfy the following requirements

$$\begin{aligned}
&\int K(x)\,dx = 1,\\
&\int xK(x)\,dx = 0,\\
&\int x^2K(x)\,dx < \infty,\\
&K(x) \geq 0 \text{ for all } x,\\
&K(x) = K(-x),
\end{aligned} \tag{3.2}$$

i.e. the kernel must be symmetric, continuous PDF with mean zero and bounded variance. In this case, K is called a kernel of order $k = 2$. In the literature, see for example [102], certain *higher-order kernels* are also being considered. In such a case, the above requirements must by rewritten in a more general form

$$\begin{aligned}
&\int K(x)\,dx = 1,\\
&\int x^jK(x)\,dx = 0 \text{ for } j = 1,\dots k-1,\\
&\int x^kK(x)\,dx \neq 0,\\
&K(x) = K(-x).
\end{aligned} \tag{3.3}$$

Note that the symmetry of K implies that k is an *even number*. The problem with the higher-order cases is that the requirement stating that $K(x) \geq 0$ for all x does not hold and some negative contributions of the kernel can result in being the estimated density not a density itself. This book deals only with the kernels of order $k = 2$.

Table 3.1 contains the most popular kernels with their shapes shown in Fig. 3.1. These are all univariate formulas but the multivariate extensions are obvious. Note also that the Uniform, Epanechnikov, Biweight and Triweight kernels are particular cases of the general kernel formula (for $p = 1, 2, 3, 4$ respectively) [56]

$$K(x, p) = \left(2^{2p+1}B(p+1, p+1)\right)^{-1}(1 - x^2)\mathbb{1}_{\{|x|<1\}}, \tag{3.4}$$

Table 3.1 Popular univariate kernel types

Kernel name	$K(x)$
Normal	$(2\pi)^{-1/2}exp(1/2\, x^2)$
Epanechnikov	$(3/4)(1-x^2)\ \mathbb{1}_{\{\lvert x\rvert<1\}}$
Uniform (Box)	$1/2\ \mathbb{1}_{\{\lvert x\rvert<1\}}$
Biweight	$(15/16)(1-x^2)^2\ \mathbb{1}_{\{\lvert x\rvert<1\}}$
Triweight	$(35/32)(1-x^2)^3\ \mathbb{1}_{\{\lvert x\rvert<1\}}$
Triangular	$(1-\lvert x\rvert)\ \mathbb{1}_{\{\lvert x\rvert<1\}}$

where $B(\cdot,\cdot)$ is the beta function and $\mathbb{1}_{\{x\in A\}}$ is the indicator function, that is $\mathbb{1}_{\{x\in A\}} = 1$ for $x \in A$ and $\mathbb{1}_{\{x\in A\}} = 0$ for $x \notin A$. These kernels are symmetric beta densities on the interval $[-1, 1]$.

3.3 Definitions

3.3.1 Univariate Case

The univariate kernel density estimator (note that—depending on the context—the KDE abbreviation can refer to kernel density *estimation* or kernel density *estimator*) for a random sample $X_1, X_2, \ldots X_n$ drawn from a common and usually unknown density f is given by

$$\hat{f}(x,h) = \frac{1}{nh}\sum_{i=1}^{n} K\left(\frac{x-X_i}{h}\right),$$
$$= \frac{1}{n}\sum_{i=1}^{n} K_h\left(x-X_i\right), \tag{3.5}$$

where

$$K_h(u) = h^{-1}K\left(h^{-1}u\right), \tag{3.6}$$

and h is the bandwidth, which is a positive real number. The scaled (K_h) and unscaled (K) kernels are related in (3.6). In most cases the kernel has the form of the standard univariate normal density, see Table 3.1. The above defined estimator is also known as *Parzen-Rosenblatt estimator*, named after its two inventors [130, 146].

The bandwidth h has the interpretation of the standard deviation σ in the normal distribution given by the very well known formula

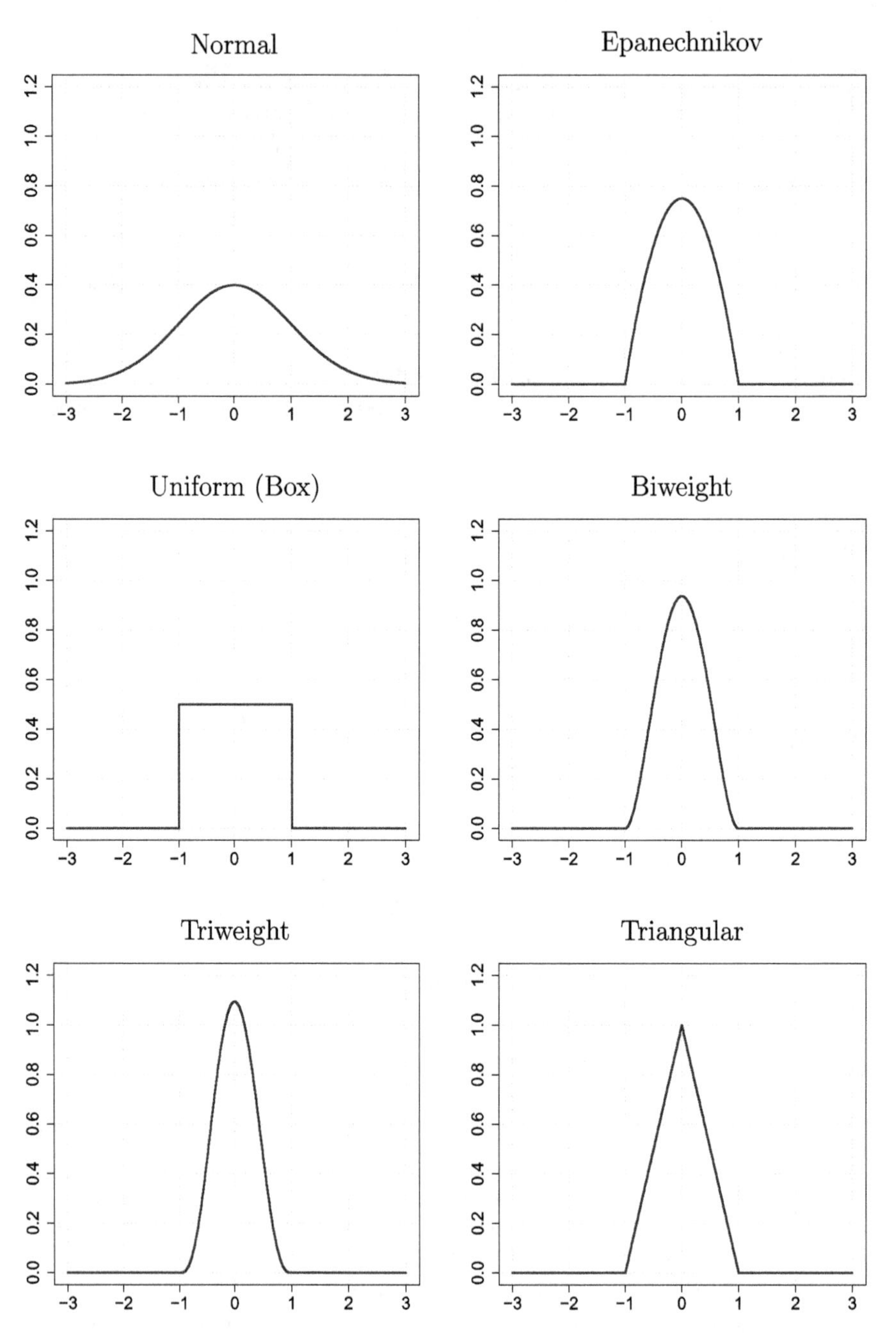

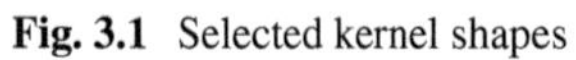

Fig. 3.1 Selected kernel shapes

$$\mathcal{N}(x; \mu, \sigma) = \frac{1}{\sqrt{(2\pi)}\sigma} \exp\left(-\frac{(x-\mu)^2}{2\sigma^2}\right). \tag{3.7}$$

The kernels K usually have the form of the standard normal distribution $\mathcal{N}(0, 1)$ given by (3.1), so

$$K\left(\frac{x - X_i}{h}\right) = \frac{1}{\sqrt{(2\pi)}} \exp\left(-\frac{(x - X_i)^2}{2h^2}\right), \tag{3.8}$$

and finally

$$\begin{aligned} \hat{f}(x, h) &= \frac{1}{nh} \sum_{i=1}^{n} \frac{1}{\sqrt{(2\pi)}} \exp\left(-\frac{(x - X_i)^2}{2h^2}\right) \\ &= \frac{1}{n} \sum_{i=1}^{n} \frac{1}{\sqrt{(2\pi)}h} \exp\left(-\frac{(x - X_i)^2}{2h^2}\right). \end{aligned} \tag{3.9}$$

Comparing (3.7) with (3.9), it should be clear that h can be interpreted as the standard deviation and obviously it must be a real positive number. Moreover, X_i data points can be interpreted as the mean values of the normal distribution. Given all that, it is evident that KDE can be understood as a *weighted sum of density 'bumps'* that are centered at each data point X_i. The shape of the bumps is determined by the choice of the kernel function (see Table 3.1). The width of the bumps is determined by the bandwidth h, that is

$$\hat{f}(x, h) = \frac{1}{n} \sum_{i=1}^{n} \mathcal{N}(x; X_i, h^2). \tag{3.10}$$

As an example of how KDE works consider a toy univariate dataset of seven data points:

$$X_1 = 0,\ X_2 = 1,\ X_3 = 1.1,\ X_4 = 1.5,\ X_5 = 1.9,\ X_6 = 3.9,\ X_7 = 4.5. \tag{3.11}$$

Four different KDEs based on these data are depicted in Fig. 3.2. It is easy to notice how the bandwidth h influences the shape of the KDE curve. Solid lines show the estimated PDFs, while dashed lines show the shapes of individual kernel functions K (Gaussians). The small filled black dots represent the data points X_i. In Fig. 3.2a, the bandwidth h is too small compared to the optimal value, producing the *undersmoothed* curve (it contains too many spurious bumps). In this example, the optimal bandwidth is about $h = 0.8$. In Fig. 3.2c and d, the bandwidth is too big, producing the *oversmoothed* curves (it obscures much of the underlying structure of the input data).

In general, estimating the optimal bandwidth is not a trivial task. The literature describing this problem is extensive and it includes a number of positions that have

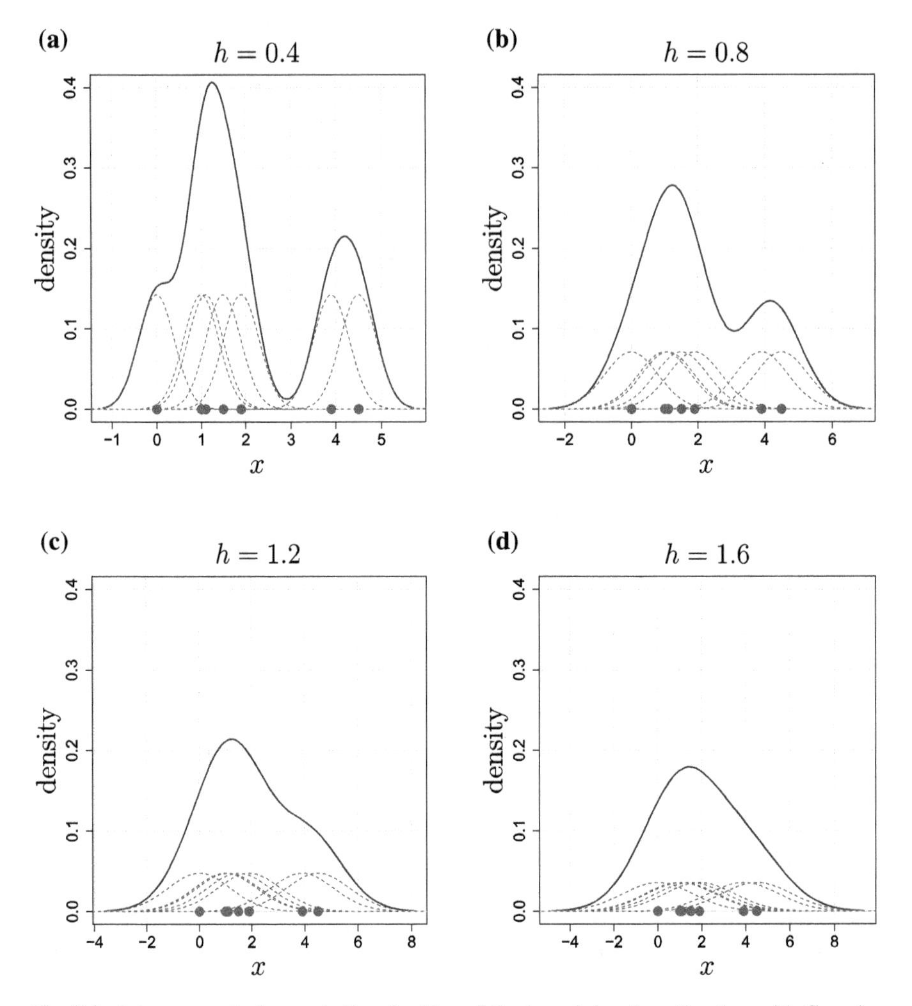

Fig. 3.2 A toy example demonstrating the idea of the kernel density estimation with Gaussian kernels

become classic, see e.g. [13, 84, 158, 168]. We describe this problem in more detail in Chap. 4. Let us just say here that usually, it is not possible to provide a single best optimal bandwidth. Typically, different methods return slightly different results (and all of them are optimal according to the precisely defined criteria used within each of the methods).

The Parzen window density estimator presented in Sect. 2.5 can be thought of as the sum of boxes generated by the kernel function (2.13) and centered at the observations (see Fig. 2.7). Here, the smooth estimate based on Gaussian kernels is the sum of 'bumps' centered at the data points.

Consider another univariate example and let X_i be a sample drawn from a mixture of two Gaussians given by

$$f(x) = \frac{7}{10}\mathcal{N}(x;\ 4,\ 0.8) + \frac{3}{10}\mathcal{N}(x;\ 6,\ 0.5)\,. \tag{3.12}$$

Three different KDEs based on data drawn from (3.12) with sample size $n = 100$ are depicted in Fig. 3.3. In this example, the optimal bandwidth $h = 0.338$ and this value was calculated using the univariate plug-in method (see Chap. 4). The true density is plotted with the dashed line. The left lower estimate is obviously the most similar to this true density.

The next example demonstrates how the kernel type (see Table 3.1) affects the density estimates. It is a well known fact that the Epanechnikov kernel is the most efficient one (see for example [188]). The efficiency is proportional to the sample size necessary to obtain the same estimation error. It is also well-known that the efficiency differences between various kernel types are not very large. However, the normal kernel generates the smoothest density curves and this is the reason why this type of kernel is most often used in practice. Figure 3.4 shows the KDEs of the true density given by (3.12). Six kernels shown in Table 3.1 were used. It is easily seen that the normal kernel gives the smoothest curve.

We emphasize that the notion of the so-called *canonical kernels* [115, 188, 191] was not used in this numerical example. Canonical kernels are useful for a pictorial comparison of the density estimates based on different kernel shapes. Such kernels are rescaled versions of non-canonical kernels and are defined in such a way that a particular single choice of bandwidth gives roughly the same amount of smoothing. Figure 3.5 is the replication of Fig. 3.4, except that now the canonical kernels have been used. The individual bandwidths were rescaled in the following way (for details, see for example [84] and [188, Sect. 2.7]). The base bandwidth (h_{base}) was calculated with the univariate plug-in method (see Chap. 4). In the next step, this base bandwidth was multiplied by an adjustment factor known as the *canonical bandwidth* corresponding to the kernel function K. The canonical bandwidths are calculated using the following scaling factor

$$\delta_0(K) = \left(\frac{R(K)}{\mu_2(K)^2}\right)^{1/5}, \tag{3.13}$$

and then

$$\begin{aligned}
h_{\text{normal}} &= \left(\frac{1}{4\pi}\right)^{1/10} h_{\text{base}} \approx 0.7764\, h_{\text{base}} \\
h_{\text{epanechnikov}} &= (15)^{1/5}\, h_{\text{base}} \approx 1.7188\, h_{\text{base}} \\
h_{\text{uniform}} &= \left(\frac{9}{2}\right)^{1/5} h_{\text{base}} \approx 1.3510\, h_{\text{base}} \\
h_{\text{biweight}} &= (35)^{1/5}\, h_{\text{base}} \approx 2.0362\, h_{\text{base}}
\end{aligned}$$

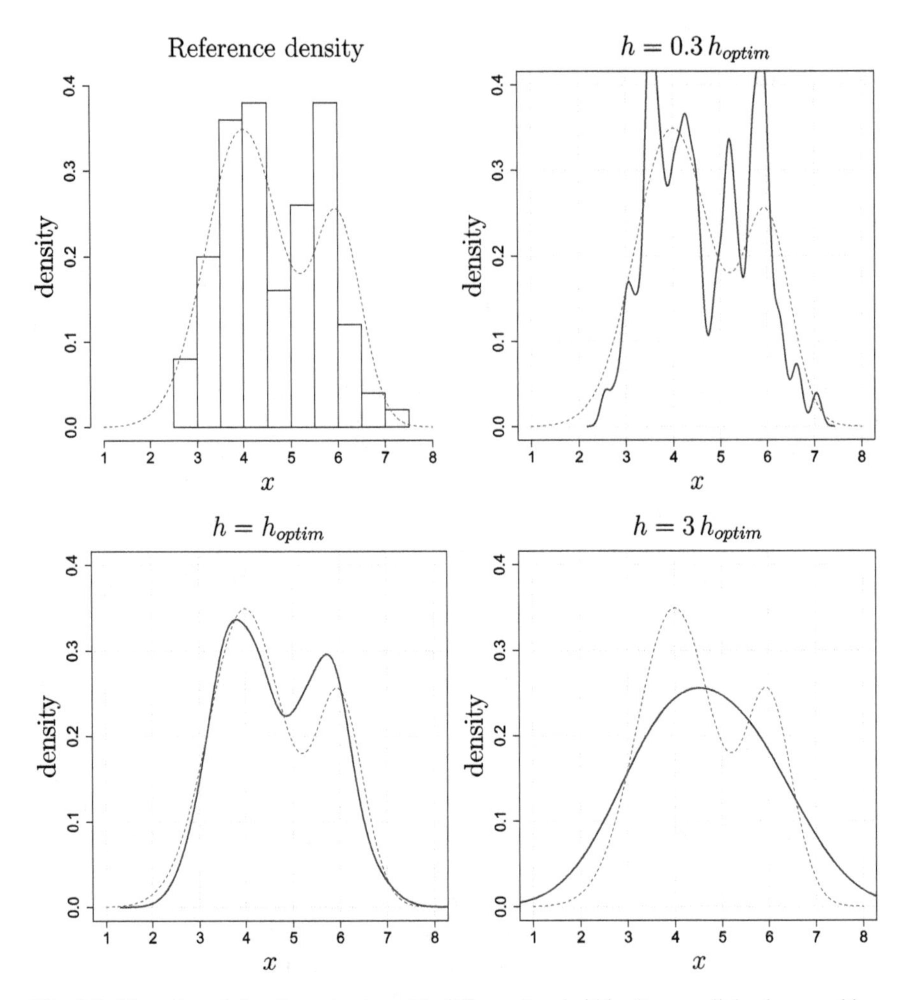

Fig. 3.3 Three kernel density estimates with different bandwidths (too small (undersmoothing; small bias but large variability), optimal, and too big (oversmoothing; small variability but large bias)). The true density curve is plotted as the dashed line

$$h_{\text{triweight}} = \left(\frac{9450}{143}\right)^{1/5} h_{\text{base}} \approx 2.3122\, h_{\text{base}}$$
$$h_{\text{triangular}} = (24)^{1/5}\, h_{\text{base}} \approx 1.8882\, h_{\text{base}}. \tag{3.14}$$

Figure 3.5 confirms that the choice of bandwidth h is much more important than the choice of kernel K. Only the uniform kernel generates a jaggy curve. Other kernels generate very similar curves, which are also sufficiently smooth.

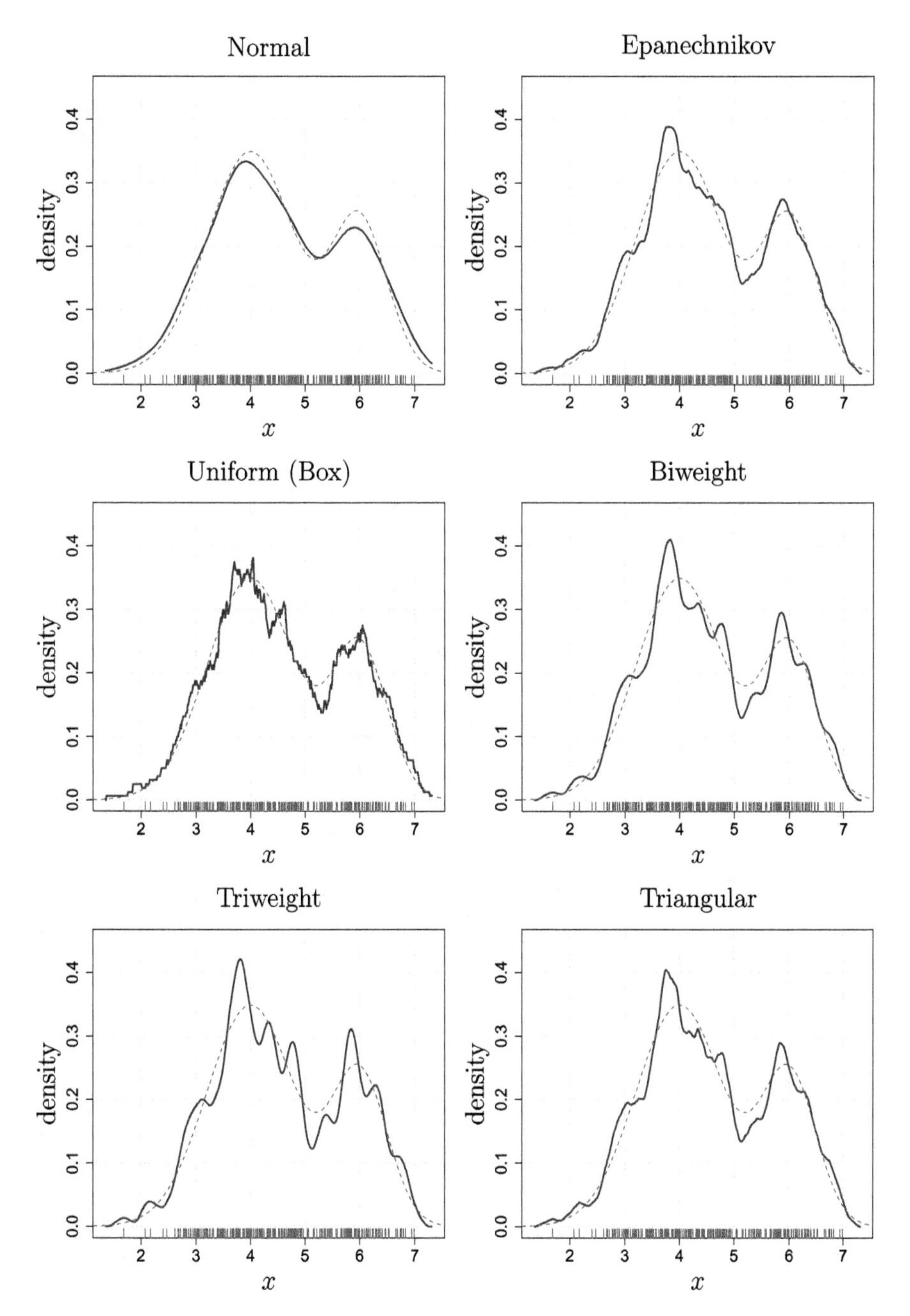

Fig. 3.4 Six different kernel types and theirs KDEs. The true density curve is plotted in the dashed line. The bandwidth $h = 0.32$ is fixed for every kernel type

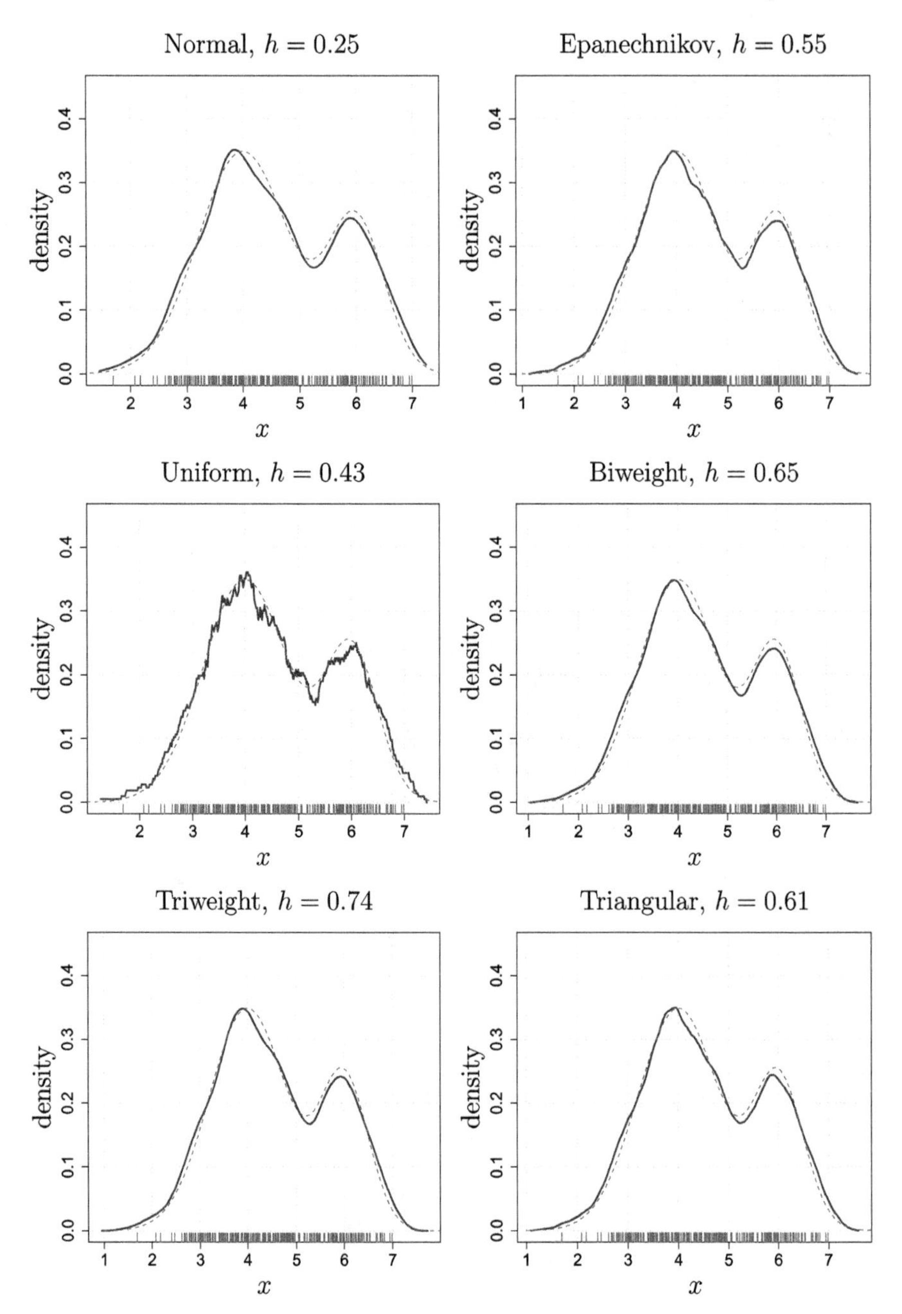

Fig. 3.5 Six different kernel types and their KDEs. The true density curve is plotted in the dashed line. The bandwidths are different for each type with their value provided in the description of the plots

3.3.2 *Multivariate Case*

A general form of the d-dimensional multivariate kernel density estimator is

$$\hat{f}(\boldsymbol{x}, \mathbf{H}) = n^{-1} \sum_{i=1}^{n} |\mathbf{H}|^{-1/2} K \left(\mathbf{H}^{-1/2} (\boldsymbol{x} - \boldsymbol{X}_i) \right)$$
$$= n^{-1} \sum_{i=1}^{n} K_{\mathbf{H}} (\boldsymbol{x} - \boldsymbol{X}_i), \qquad (3.15)$$

where

$$K_{\mathbf{H}}(\boldsymbol{x}) = |\mathbf{H}|^{-1/2} K \left(\mathbf{H}^{-1/2} \boldsymbol{x} \right), \qquad (3.16)$$

and $\mathbf{H}$ is the $d \times d$ *bandwidth matrix* or *smoothing matrix*, which is non-random *symmetric* and *positive definite*, and d is the problem dimensionality; $\boldsymbol{x} = (x_1, x_2, \ldots, x_d)^T$ and $\boldsymbol{X}_i = (X_{i1}, X_{i2}, \ldots, X_{id})^T$, $i = 1, 2, \ldots, n$ is a sequence of independent identically distributed (*iid*) d-variate random variables drawn from a (usually unknown) density f. Here K and $K_{\mathbf{H}}$ are the *unscaled* and *scaled* kernels, respectively. In most cases, the kernel has the form of a standard multivariate normal density given by the following formula

$$K(\boldsymbol{u}) = (2\pi)^{-d/2} \exp \left(-\frac{1}{2} \boldsymbol{u}^T \boldsymbol{u} \right). \qquad (3.17)$$

Multivariate extensions of the popular kernel types shown in Table 3.1 can also be used. Similarly to the case of the formula (3.10), the multivariate KDE can be viewed as a *weighted sum of density 'bumps'* that are centered at each data point $\boldsymbol{X}_i$, that is

$$\hat{f}(\boldsymbol{x}, \mathbf{H}) = \frac{1}{n} \sum_{i=1}^{n} \mathcal{N}(\boldsymbol{x}; \boldsymbol{X}_i, \mathbf{H}). \qquad (3.18)$$

Note that the notation used in (3.16) is not a direct extension of the univariate notation in (3.6), since in the one-dimensional case the bandwidth is $\mathbf{H} = h^2$, hence, in the present case, 'squared bandwidths' are used.

The most general form (3.15) is much more computationally and mathematically complex, compared to its univariate counterpart (3.5). In this context, the most important problem is to select the optimal bandwidth matrix. The bandwidth matrix $\mathbf{H}$ affects both the *shape* and the *orientation* of the kernels. This behavior is demonstrated in a bivariate toy example consisting of five data points

$$\boldsymbol{X}_1 = (7, 3), \ \boldsymbol{X}_2 = (2, 4), \ \boldsymbol{X}_3 = (4, 4), \ \boldsymbol{X}_4 = (5, 2), \ \boldsymbol{X}_5 = (6, 7), \qquad (3.19)$$

and three bandwidth matrices

$$\mathbf{H}_1 = \begin{bmatrix} 1 & 0.7 \\ 0.7 & 1 \end{bmatrix}, \quad \mathbf{H}_2 = \begin{bmatrix} 1 & 0 \\ 0 & 1 \end{bmatrix}, \quad \mathbf{H}_3 = \begin{bmatrix} 3.14 & -0.054 \\ -0.054 & 2.08 \end{bmatrix}. \tag{3.20}$$

The relevant KDEs are depicted in Fig. 3.6. The left panel shows the individual kernels, centered at each data point, while the right panel shows the density estimates. The bandwidth $\mathbf{H}_3$ is the optimal one when using the multivariate plug-in method described in Chap. 4. Comparing bandwidth values with the corresponding kernel shapes (Fig. 3.6a,c,e) it not difficult to notice how the individual values affect the shape and orientation of the kernels. The bandwidth matrix $\mathbf{H}_1$ generates kernels with the north-east orientation, the unit bandwidth matrix $\mathbf{H}_2$ generates circular kernels and the bandwidth matrix $\mathbf{H}_3$ generates slightly south-east oriented kernels (the off-diagonal entries of this matrix are very small compared to the diagonal entries, so the kernel deviation from the $x - y$ axis is minimal).

When dealing with a bivariate case, contour plots are very useful in a more detailed inspection of the KDE results. Perspective plots can also be useful but can only provide a rough visual aid. For an example, consider Fig. 3.7, which shows both plots. This example is based on a real bivariate dataset (*faithful*, from the *datasets R* package) with 272 observations on 2 variables.

A graphical KDE presentation in cases of 3-dimensional datasets is also possible, see [160], and an appropriate example (see Fig. 3.8) shows a 3-dimensional KDE for the classical *iris* dataset [4] (available for example in the *datasets R* package). Three (from the total of four) variables were used in this example. This plot combines both the 3D scatterplot and the 3D contour plot that depicts the density values. Darker fragments represent regions with higher density values. This technique for visualization of three dimensional data was proposed in [51]. In this example, the optimal bandwidth (calculated using the multivariate plug-in method described in Chap. 4) is

$$\mathbf{H} = \begin{bmatrix} 0.026 & -0.013 & -0.003 \\ -0.013 & 0.212 & 0.087 \\ -0.003 & 0.087 & 0.041 \end{bmatrix}. \tag{3.21}$$

A graphical KDE presentation for more that four dimensions is of course not directly attainable. However, a technique known as 'slicing' is a potential workaround (plotting of three-dimensional subset with the other parameters remaining constant).

In the literature, the bandwidth matrix is defined on three levels of complexity. Let $\mathcal{F}$ be the class of symmetric, positive definite $d \times d$ matrices. The simplest case is when a positive constant scalar multiplies the identity matrix, that is, $\mathbf{H} \in \mathcal{S}$, where $\mathcal{S} = \{h^2 \mathbf{I}_d : h > 0\}$. Another level of sophistication is when $\mathbf{H} \in \mathcal{D}$, where $\mathcal{D} \subseteq \mathcal{F}$ is the subclass of diagonal positive definite $d \times d$ matrices. For $\mathbf{H} \in \mathcal{D}$, $D = \mathrm{diag}(h_1^2, \ldots h_d^2)$. These two forms are often called *constrained*. In the most general form, the bandwidth is *unconstrained*, that is with $\mathbf{H} \in \mathcal{F}$. Below, only the bivariate case is shown, but the generalization to a multivariate case is easy

1. Positive constant times the identity matrices: $\mathbf{H}_{\mathcal{S}} = h^2 \mathbf{I} = \begin{bmatrix} h^2 & 0 \\ 0 & h^2 \end{bmatrix}$,

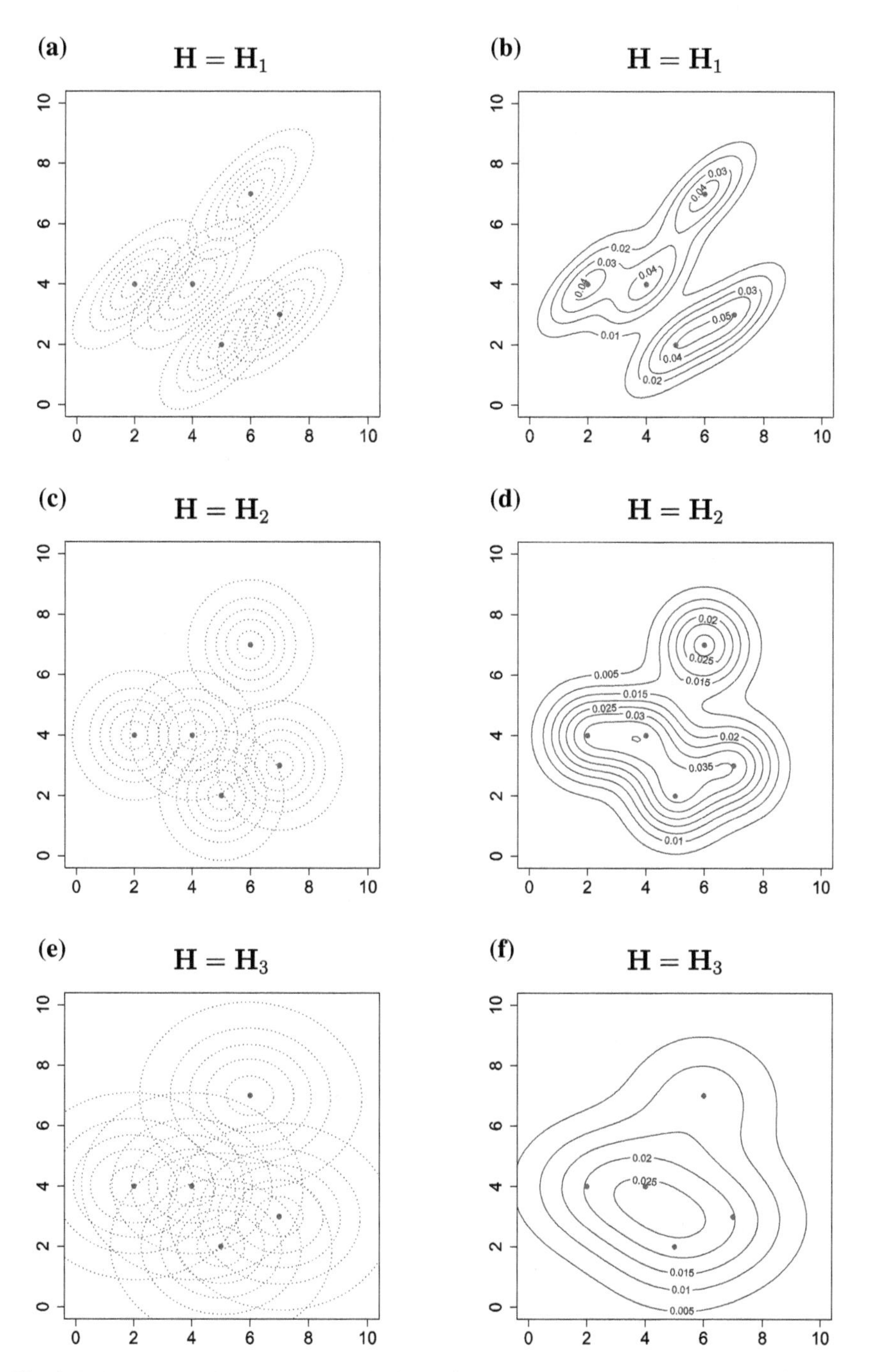

Fig. 3.6 A toy example demonstrates the idea of the kernel density estimation with Gaussian kernels. Solid lines—kernel density estimates, dashed lines—individual kernels centered at each data point

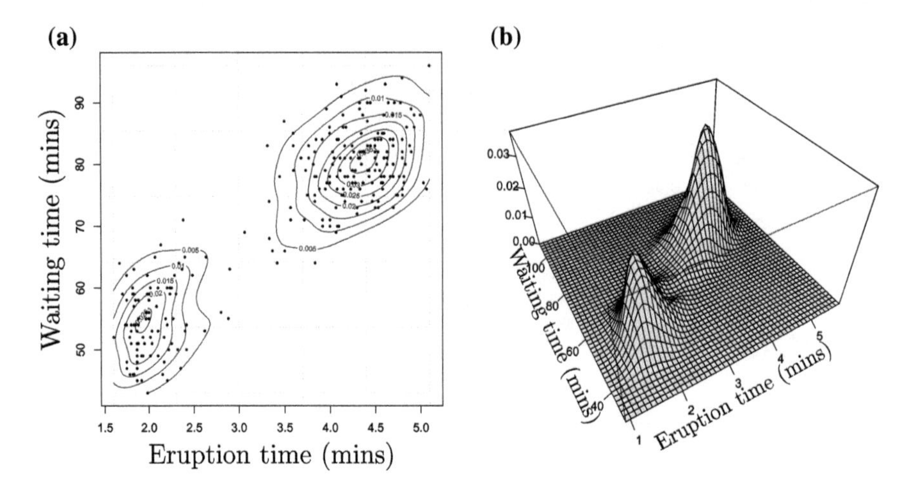

Fig. 3.7 An illustrative example of contour-type and perspective-type presentations of a bivariate KDE. A sample 2D dataset *faithful* was used in this example

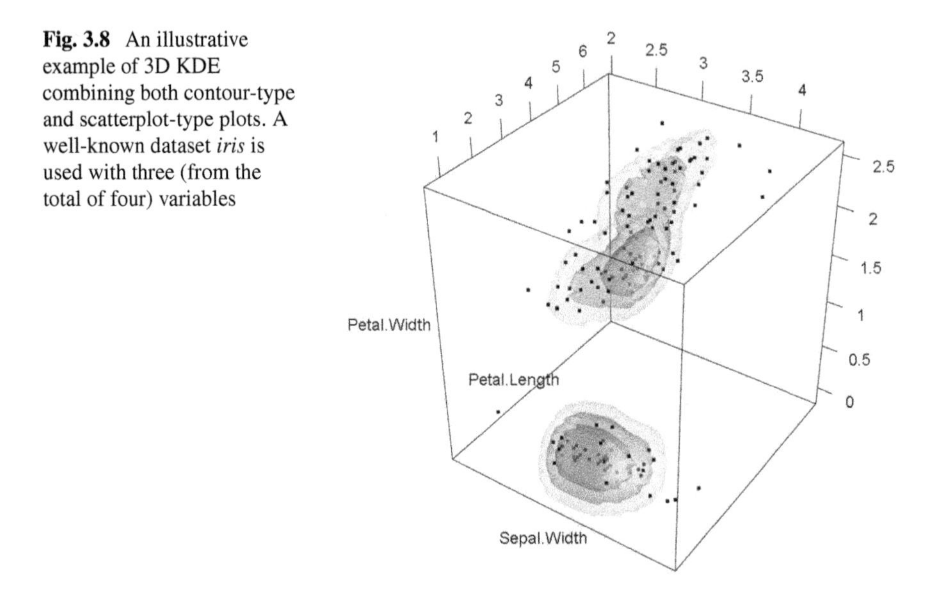

Fig. 3.8 An illustrative example of 3D KDE combining both contour-type and scatterplot-type plots. A well-known dataset *iris* is used with three (from the total of four) variables

2. Diagonal positive definite matrices: $\mathbf{H}_{\mathcal{D}} = \text{diag}\mathbf{H} = \begin{bmatrix} h_1^2 & 0 \\ 0 & h_2^2 \end{bmatrix}$,

3. Full symmetric positive definite matrices: $\mathbf{H}_{\mathcal{F}} = \begin{bmatrix} h_1^2 & h_{12} \\ h_{12} & h_2^2 \end{bmatrix}$.

For $\mathbf{H} \in \mathcal{D}$ and $\mathbf{H} \in \mathcal{S}$ the kernel estimators can be simplified, respectively as

$$\hat{f}(\boldsymbol{x}, \mathbf{H}) = \frac{1}{n} \left(\prod_{l=1}^{d} h_l \right)^{-1} \sum_{i=1}^{n} K \left(\frac{x_1 - X_{i1}}{h_1}, \ldots, \frac{x_d - X_{id}}{h_d} \right), \tag{3.22}$$

and

$$\hat{f}(\boldsymbol{x}, h) = \frac{1}{nh^d} \sum_{i=1}^{n} K \left(\frac{\boldsymbol{x} - \boldsymbol{X}_i}{h} \right). \tag{3.23}$$

The kernel function K is almost always taken to be a d-variate probability density function (see Table 3.1). There are two popular techniques used to generate multivariate kernels from a *symmetric univariate kernel* $\mathcal{K}$ (note that a different symbol was used here, to differentiate univariate ($\mathcal{K}$) and multivariate (K) kernels). The first is called the *product kernel* and is defined as

$$K^{\mathrm{P}}(\boldsymbol{x}) = \prod_{l=1}^{d} \mathcal{K}(x_i), \tag{3.24}$$

and the second is called the *radial kernel* or *radially symmetric kernel* and is defined as

$$K^{\mathrm{R}}(\boldsymbol{x}) = c_{\mathcal{K},d}\, \mathcal{K}\{(\boldsymbol{x}^T \boldsymbol{x})^{1/2}\}, \tag{3.25}$$

where the constant $c_{\mathcal{K},d}$ is chosen in such a way that the kernel K^{R} satisfies the first condition in (3.2). Kernels of the form (3.25) use observations from a circle around $\boldsymbol{x}$ to estimate PDF at $\boldsymbol{x}$. The name comes from the fact that $K^{\mathrm{R}}(\boldsymbol{u})$ has the same value for all $\boldsymbol{u}$ on a sphere around zero. Figure 3.9 shows contours for the product K^{P} and radially symmetric K^{R} kernels. In general, these are different, with the exception being the normal density. It is obvious that the radial kernels are the only available option when working with the most general form of KDE described by (3.15).

From the theoretical point of view, the most general form of KDE as given by (3.15) should provide the best density estimators, but the price payed for their accuracy is such that they are the most complex and thus their formal analysis is the most difficult. On the other hand, the use of a single smoothing parameter h as in (3.23) is a very attractive idea for practical purposes as only one parameter (single scalar h) must be estimated. However, the basic requirement here is that the spread of data (measured by the variance) in every dimension should be at least approximately the same. In many real datasets this is not the case. In such situations, it is worth to consider a data pre-scaling procedure, that is a way of *linearly transforming* the input data to obtain unit covariance matrix. This can be done by the so called *whitening transformation* (the transformation is called 'whitening' because it changes the input vector into a white noise vector). Next, the pre-scaled data are smoothed using a radially symmetric

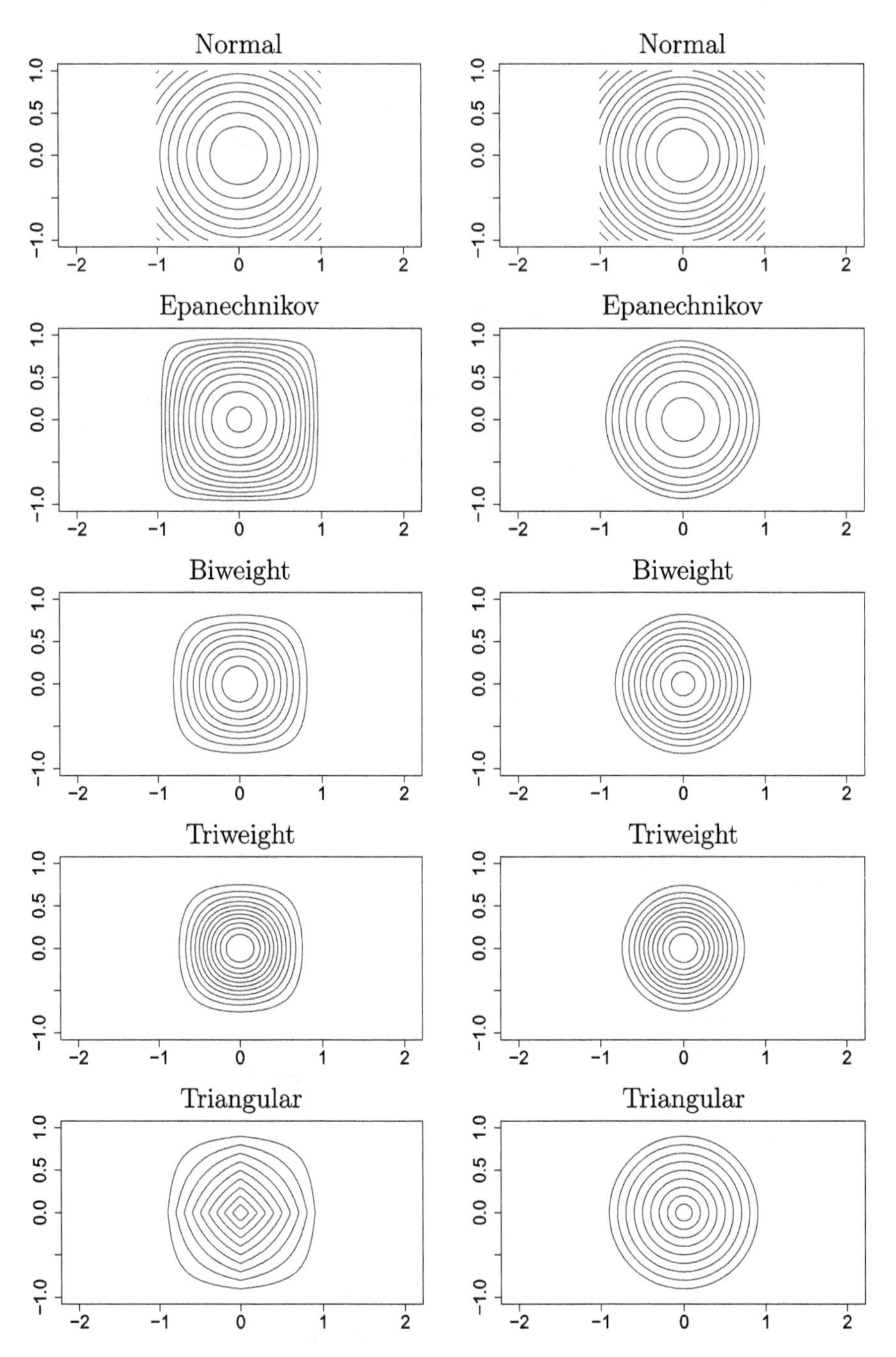

Fig. 3.9 Contours from product (left column) and radially symmetric (right column) kernels with equal bandwidths in each direction, i.e. $\mathbf{H} = (h_1, h_2)^T = (1, 1)^T$

kernel, and finally the result is transformed back [168]. The above procedure is equivalent to using a generalized radially symmetric kernel of the following form

$$K(\boldsymbol{u}) = (2\pi)^{-d/2}|\Sigma|^{-1/2}exp\left(-\frac{1}{2}\boldsymbol{u}^T\Sigma^{-1}\boldsymbol{u}\right), \tag{3.26}$$

where Σ is the input data covariance matrix. The positive effect of using (3.26) is that the kernel contours are better adapted to the overall data shape. The complete procedure is then as follows

1. Apply the whitening transformation

$$\boldsymbol{Y} = \Sigma^{-1/2}\boldsymbol{X}. \tag{3.27}$$

2. Smooth the pre-scaled data

$$\hat{f}(\boldsymbol{y}, h) = \frac{1}{nh^d}\sum_{i=1}^{n} K\left(\frac{\boldsymbol{y} - \boldsymbol{Y}_i}{h}\right). \tag{3.28}$$

3. Transform back

$$\hat{f}(\boldsymbol{x}, h) = \hat{f}(\boldsymbol{y}, h)|\Sigma|^{-1/2}. \tag{3.29}$$

The notion of the linear transformation is depicted in Fig. 3.10. Consider a sample random variable $\mathbf{X}$, having the following covariance matrix

$$\Sigma = \begin{bmatrix} 4 & 1.2 \\ 1.2 & 1 \end{bmatrix}. \tag{3.30}$$

We can encounter three shapes of kernels. The kernels are shaped as circles, if Σ is set to be the identity matrix, see Fig. 3.10a. It can be easily seen that the shapes do not follow the north-east spread of the data. If Σ is set to be the diagonal matrix (the off-diagonals in (3.30) have been zeroed), the kernels are shaped as ellipses and also do not follow the data orientation in a satisfactory way. Finally, if Σ is the full symmetric positive definite matrix (3.30), the kernels start following the shape of the data set itself.

Note that the above procedure can not be used blindly for every dataset. The following example (taken from [19]) shows that it is easy to generate examples where even after pre-scaling, the output dataset ($\boldsymbol{Y}$) does not seem to follow a spherically-symmetric distribution. Figure 3.11 presents the two scatterplots for samples of size $n = 5000$. The first one is unimodal (Fig. 3.11a) and the second is asymmetric bimodal (Fig. 3.11c). After the whitening transformation (3.27), the two output datasets $\boldsymbol{Y}$ have the identity covariance matrices. Yet, a spherically symmetric distribution in Fig. 3.11d is not obtained at all.

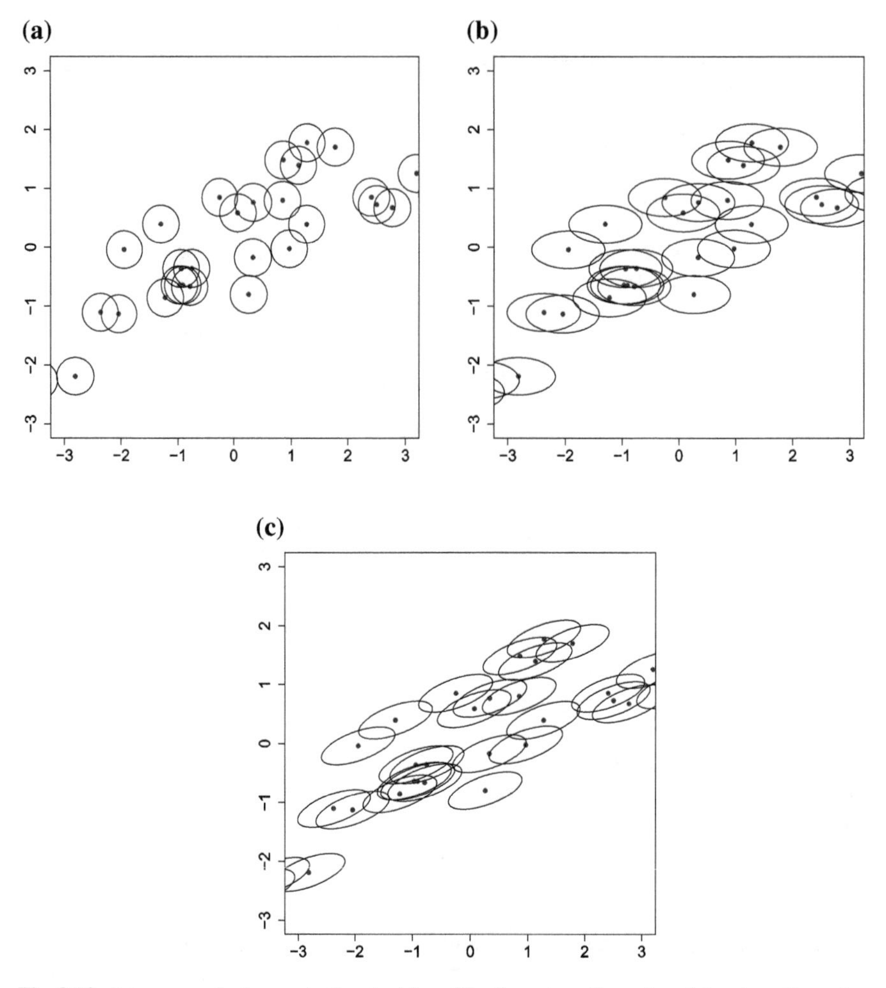

Fig. 3.10 A toy example demonstrating the idea of the linear transformation: (**a**) no transformation used, (**b**) diagonal covariance matrix used, (**c**) full covariance matrix used

3.4 Performance Criteria

3.4.1 Univariate Case

The performance analysis of the kernel density estimation requires the specification of appropriate error criteria for measuring the closeness of $\hat{f}$ to its target density f. Arguably, one the most commonly used *local* error criteria includes the *Mean Squared Error* (MSE) that measures the closeness of an estimator $\hat{\theta}$ to its (often unknown) target parameter θ

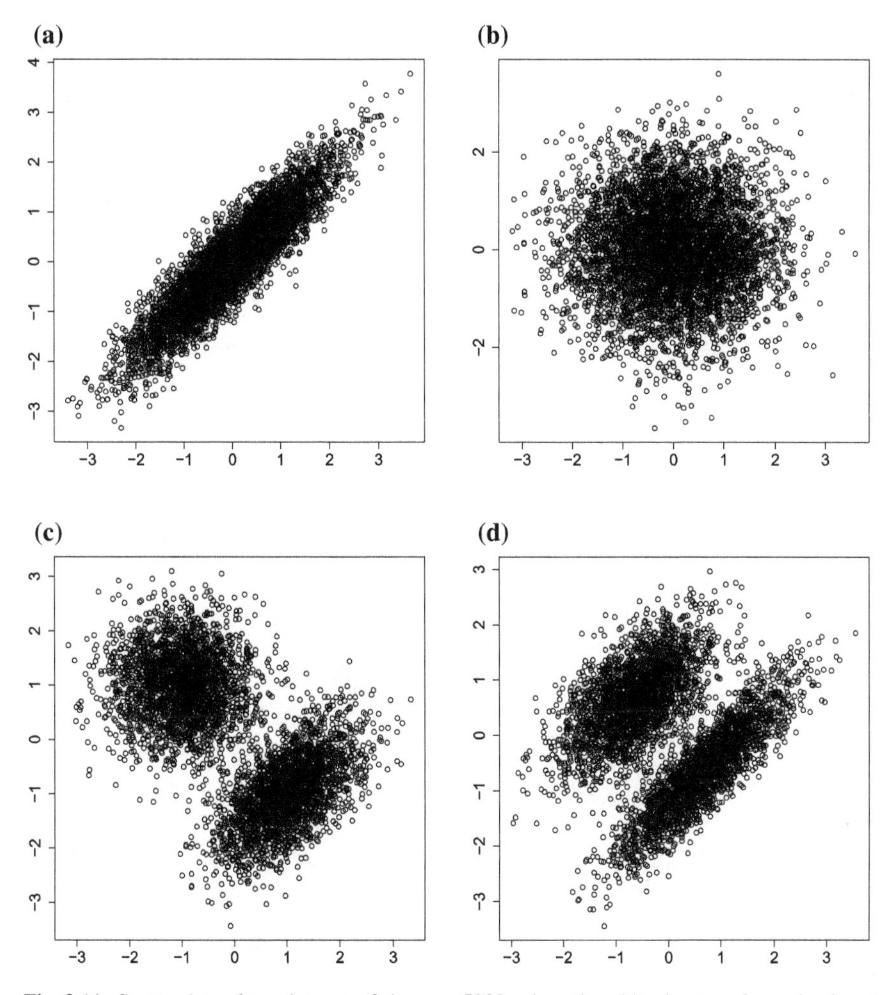

Fig. 3.11 Scatterplots of two datasets of size $n = 5000$, where the whitening transformation is not enough to get spherically-symmetric distributions

$$\mathrm{MSE}(\hat{\theta}) = \mathrm{E}(\hat{\theta} - \theta)^2, \tag{3.31}$$

which can be decomposed into variance and squared bias

$$\mathrm{MSE}(\hat{\theta}) = \mathrm{Var}\,\hat{\theta} + (\mathrm{E}\,\hat{\theta} - \theta)^2. \tag{3.32}$$

In the case of the kernel density $\hat{f}(x; h)$ at some point $x \in \mathrm{R}$ (note that here the notation highlights the explicit dependence on h) the above criterion is obviously

$$\mathrm{MSE}\,\hat{f}(x; h) = \mathrm{E}(\hat{f}(x; h) - f(x))^2$$

$$= \text{Var}\, \hat{f}(x; h) - (\text{E}\, \hat{f}(x; h) - f(x))^2. \tag{3.33}$$

The above is a local error at a fixed point x. In practice, it is usually desirable to consider a *global* error criterion that measures the distance between $\hat{f}$ and f over the entire domain. One such criterion is the *Integrated Squared Error* (ISE). Note that the notation $\hat{f}(\cdot; h)$ is used to indicate that now it is the global error that is measured instead of an error at a particular point x

$$\text{ISE}\, \hat{f}(\cdot; h) = \int_{-\infty}^{+\infty} (\hat{f}(x; h) - f(x))^2\, dx. \tag{3.34}$$

Thus, the ISE summarizes the performance of $\hat{f}$ as a function of the observed data (it is a stochastic variable). Therefore, it is more appropriate to analyze the expected value of this stochastic entity, that is the *Mean Integrated Squared Error* (MISE)

$$\text{MISE}\, \hat{f}(\cdot; h) = \text{E}(\text{ISE}\, \hat{f}(\cdot; h)) = \text{E} \int_{-\infty}^{+\infty} (\hat{f}(x; h) - f(x))^2\, dx, \tag{3.35}$$

where the expectation is taken with respect to the distribution f. Thus, MISE can be viewed as the average value of the global ISE measure of error with respect to the sampling density. Moreover, by changing the integral and expectation operators we have

$$\begin{aligned} \text{MISE}\, \hat{f}(\cdot; h) &= \int_{-\infty}^{+\infty} \text{E}(\hat{f}(x; h) - f(x))^2\, dx = \int_{-\infty}^{+\infty} \text{MSE}\, \hat{f}(x; h)\, dx \\ &= \int_{-\infty}^{+\infty} \text{Var}\, \hat{f}(x; h)\, dx + \int_{-\infty}^{+\infty} \text{Bias}^2\, \hat{f}(x; h)\, dx. \end{aligned} \tag{3.36}$$

The bandwidth parameter h controls the smoothness of the density estimate, which was for example demonstrated in Fig. 3.3. Thus, one needs to find the following

$$h_{\text{MISE}} = \underset{h \in \mathbb{R}^+}{\text{argmin}}\ \text{MISE}\, \hat{f}(\cdot, h). \tag{3.37}$$

Finding h_{MISE} is in general not an easy task. The most important methods for doing so are briefly presented in Sect. 4.2.

What is also very important to note is that the MISE does not have a closed form, as f is in general unknown (except for a situation where f is a normal mixture density and K is the normal kernel [188, Sect. 2.6]). The usual approach is to find a *large sample approximation* of the MISE using the *Taylor series expansion* technique. We omit the details (for a comprehensive mathematical analysis see [188]) and present only the final results. The following form of the *Asymptotic* MISE (AMISE) is derived in [188]. The expression for the bias is

$$\mathrm{E}\,\hat{f}(x;h) - f(x) = \frac{1}{2}h^2\mu_2(K)f''(x) + o(h^2), \tag{3.38}$$

and the expression for the variance is

$$\mathrm{Var}\,\hat{f}(x;h) = (nh)^{-1}h^2R(K)f(x) + o((nh)^{-1}), \tag{3.39}$$

where

$$\begin{aligned} R(K) &= \int K(x)^2\,dx, \\ \mu_2(K) &= \int x^2K(x)\,dx, \\ R(f'') &= \int f''(x)^2\,dx. \end{aligned} \tag{3.40}$$

Here f is assumed to be sufficiently smooth: its second derivative f'' is bounded, continuous, square integrable and ultimately monotone (an ultimately monotone function is one that is monotone over both $(-\infty, -M)$ and (M, ∞) for some $M > 0$) [188]. Note also that higher-order smooth derivatives are required for some methods of the optimal bandwidth selection, see Sect. 4.2.

For the normal kernel and for the univariate cases, the expressions in (3.40) become

$$\begin{aligned} R(K) &= (2\pi^{1/2})^{-1}, \\ \mu_2(K) &= 1, \\ R(f'') &= 3(8\pi^{1/2})^{-1}. \end{aligned} \tag{3.41}$$

For other kernels the appropriate values can be found in [188, Appendix B]. Adding square of (3.38) and (3.39), we obtain the following

$$\begin{aligned} &\mathrm{MSE}\,\hat{f}(x;h) = \\ &(nh)^{-1}h^2R(K)f(x) + \frac{1}{4}h^4\mu_2(K)^2f''(x)^2 + o((nh)^{-1} + h^4), \end{aligned} \tag{3.42}$$

and after performing the integration of this expression, we get

$$\mathrm{MISE}\,\hat{f}(\cdot;h) = \mathrm{AMISE}\,\hat{f}(\cdot;h) + o((nh)^{-1} + h^4), \tag{3.43}$$

where

$$\mathrm{AMISE}\,\hat{f}(\cdot;h) = (nh)^{-1}R(K) + \frac{1}{4}h^4\mu_2(K)^2R(f''). \tag{3.44}$$

The above is called the *asymptotic* MISE, since it provides a useful large sample approximation to the MISE. If $(nh)^{-1} \to 0$ and $h \to 0$ as $n \to \infty$ then $\text{MISE}\hat{f}(\cdot; h) \to 0$.

The minimum value of the AMISE is easily obtained from (3.44) by differentiating with respect to h and calculating the root of the derivative. This allows to find the optimal AMISE bandwidth with the following closed form

$$h_{\text{MISE}} \sim h_{\text{AMISE}} = \left(\frac{R(K)}{\mu_2(K)^2 R(f'')n} \right)^{1/5}. \tag{3.45}$$

Substituting (3.45) into (3.44), one obtains

$$\text{AMISE}\hat{f}(\cdot; h) = \frac{5}{4}(\mu_2(K)^2 R(K)^4 R(f''))^{1/5} n^{-4/5}, \tag{3.46}$$

and this is the smallest possible AMISE for estimating f using the kernel K. The proceeding terms $n^{-4/5}$ are constant with respect to n. In other words, the AMISE *is converging at the rate* of $n^{-4/5}$.

The concept of rate of convergence has a simple interpretation in terms of the speed at which an estimator approaches its target as the sample size grows and can be very useful for a comparison of different competing estimators [188, Sect. 2.4]. It is also worth noting that in case of the histogram, the AMISE is converging at the slower rate of $n^{-2/3}$, providing yet another reason why the kernel density estimator is superior to the histogram.

The result given by (3.45) cannot be immediately made use of, since it depends on the unknown density f. A variety of automated bandwidth selectors have been proposed where the unknown f is estimated in a number of ways; see Chap. 4 for a short review of the selectors for both univariate and multivariate cases.

From (3.44), it can be seen that the integrated squared bias is asymptotically proportional to h^4, so to decrease this part one needs to choose h as small as possible. On the other hand, the integrated variance is asymptotically proportional to $(nh)^{-1}$, so to decrease this part one needs to choose h as big as possible. This is a case of a typical *trade-off* between the variance and the bias. The optimal bandwidth should be chosen in such a way that the AMISE is minimized.

Figure 3.12 shows the MISE and the AMISE as a function of the bandwidth h for $\mathcal{N}(0, 1)$ with $n = 100$. The integrated squared bias and the integrated variance are also plotted. In this example, $h_{\text{MISE}} = 0.446$ and $h_{\text{AMISE}} = 0.421$. As n is not very big, the value of h_{MISE} differs slightly from h_{AMISE}. Of course, the two values converge as $n \to \infty$. For example, for $n = 10000$ $h_{\text{AMISE}} = 0.168$ and $h_{\text{MISE}} = 0.170$. AMISE was calculated from (3.44) taking into account (3.41).

Since ISE, MISE and AMISE have closed form expressions in case of normal mixture densities, they can be computed exactly, as it was demonstrated in [188, Sect. 2.6] and implemented in the *ks R* package (`ks::ise.mixt` function).

A similar (but rather more complicated) analysis for multivariate case is also possible, both for the full bandwidth matrix **H** and for the simplified versions, when

$\mathbf{H} = h^d\mathbf{I}$ (the class of all positive constant scalarbandwidth!positive constant scalar times the identity matrix) and $\mathbf{H} = \text{diag}\mathbf{H}$ (the class of all diagonal positive definite matrices). For further information on that topic, see e.g. [18–21, 23, 24, 46, 51–54, 188].

Rounding off this section, we should also point out that (3.44) can be rewritten in the following equivalent form

$$\text{AMISE}\,\hat{f}(\cdot;h) = (nh)^{-1}\,R(K) + \frac{1}{4}h^4\mu_2(K)^2\Psi_4, \tag{3.47}$$

and also

$$h_{\text{MISE}} \sim h_{\text{AMISE}} = \left(\frac{R(K)}{\mu_2(K)^2\Psi_4 n}\right)^{1/5}. \tag{3.48}$$

Here

$$\Psi_4 = \int_{-\infty}^{\infty} f^{(4)}(x)f(x)dx, \tag{3.49}$$

or more generally

$$\Psi_r = \int_{-\infty}^{\infty} f^{(r)}(x)f(x)dx = \text{E}\,f^{(r)}(X), \tag{3.50}$$

with r an even number. The Ψ_4 (Ψ_r) naming convention was introduced in [188, Sect. 3.5]. This form is more convenient in the context of multivariate bandwidth selection algorithms, and it also is commonly used in the literature. Its estimator is defined as follows

$$\hat{\Psi}_r(g) = n^{-1}\sum_{i=1}^{n}\hat{f}^{(r)}(X_i;g) = n^{-2}\sum_{i=1}^{n}\sum_{j=1}^{n}L_g^{(r)}(X_i - X_j), \tag{3.51}$$

where g and L are, respectively, a bandwidth and a kernel that are potentially different from h and K [2, 78].

It is worth to note that the ISE and MISE error criteria are not the only ones used in the literature. In [39] the authors analyze nonparametric density estimation using the *Mean Integrated Absolute Error* (MIAE), the so called *L1 View*, where the square in the MISE criterion is replaced by the absolute value, that is

$$\text{MIAE}\,\hat{f}(\cdot;h) = \text{E}\int_{-\infty}^{+\infty}\left|\hat{f}(x;h) - f(x)\right|\,dx. \tag{3.52}$$

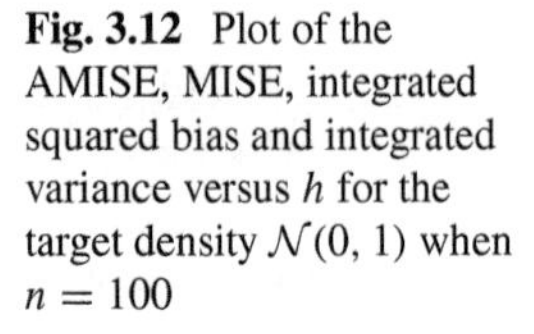

Fig. 3.12 Plot of the AMISE, MISE, integrated squared bias and integrated variance versus h for the target density $\mathcal{N}(0, 1)$ when $n = 100$

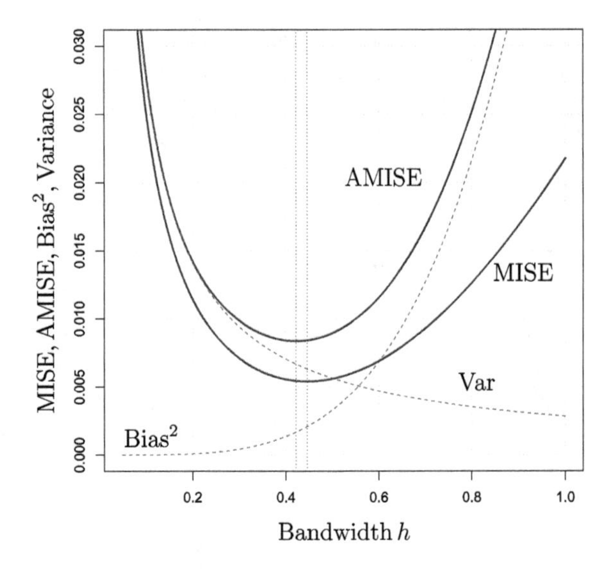

3.4.2 *Multivariate Case*

Similarly to the univariate case, most researchers use the multivariate ISE criterion as a starting point in a more complex performance analysis, that is

$$\text{ISE}\, \hat{f}(\cdot; \mathbf{H}) = \int_{\mathbb{R}^d} (\hat{f}(\boldsymbol{x}; \mathbf{H}) - f(\boldsymbol{x}))^2 \, d\boldsymbol{x}, \tag{3.53}$$

and

$$\begin{aligned} \text{MISE}\, \hat{f}(\cdot; \mathbf{H}) &= \int_{\mathbb{R}^d} \text{MSE}\, \hat{f}(\boldsymbol{x}; \mathbf{H}) \, d\boldsymbol{x} \\ &= \int_{\mathbb{R}^d} \text{Var}\, \hat{f}(\boldsymbol{x}; \mathbf{H}) \, d\boldsymbol{x} + \int_{\mathbb{R}^d} \text{Bias}^2\, \hat{f}(\boldsymbol{x}; \mathbf{H}) \, d\boldsymbol{x}. \end{aligned} \tag{3.54}$$

The bandwidth matrix $\mathbf{H}$ controls the smoothness of the density estimate. Thus, one needs to find

$$\mathbf{H}_{\text{MISE}} = \underset{\mathbf{H} \in \mathcal{F}}{\text{argmin}}\, \text{MISE}\, \hat{f}(\cdot, \mathbf{H}) \tag{3.55}$$

where $\mathcal{F}$ is the space of all symmetric, positive definite $d \times d$ matrices.

We present only the most important research outcomes, the reader is invited to consult the relevant references to fill in all the details of the derivation of the formulas presented here. The AMISE of $\hat{f}$ is given by the following equation

$$\begin{aligned}\text{AMISE}(\mathbf{H}) &\equiv \text{AMISE}\hat{f}(\cdot;\mathbf{H})\\ &= n^{-1}|\mathbf{H}|^{-1/2}R(K) + \frac{1}{4}\mu_2(K)^2(\text{vech}^T\mathbf{H})\boldsymbol{\Psi}_4(\text{vech}\mathbf{H}),\end{aligned} \tag{3.56}$$

where

$$\begin{aligned} R(K) &= \int_{\mathbb{R}^d} K(\boldsymbol{x})^2\, d\boldsymbol{x} < \infty,\\ \mu_2(K)\mathbf{I}_d &= \int_{\mathbb{R}^d} \boldsymbol{x}\boldsymbol{x}^T K(\boldsymbol{x})\, d\boldsymbol{x},\\ \mu_2(K) &< \infty,\end{aligned} \tag{3.57}$$

and vech$\mathbf{H}$ is the *vector half operator* or *half-vectorization* [91], for example

$$\text{vech}\begin{bmatrix} a & b \\ b & d\end{bmatrix} = \begin{bmatrix} a \\ b \\ d\end{bmatrix}. \tag{3.58}$$

The matrix $\boldsymbol{\Psi}_4$ is of dimension $\frac{1}{2}d(d+1) \times \frac{1}{2}d(d+1)$ and is given by

$$\boldsymbol{\Psi}_4 = \int_{\mathbb{R}^d} \text{vech}\{2\mathsf{H}f(\boldsymbol{x}) - \text{dg}\mathsf{H}f(\boldsymbol{x})\}\text{vech}^T\{2\mathsf{H}f(\boldsymbol{x}) - \text{diag}\mathsf{H}f(\boldsymbol{x})\}d\boldsymbol{x}. \tag{3.59}$$

$\mathsf{H}f = \delta^2 f/(\delta\boldsymbol{x}\delta\boldsymbol{x}^T) \in \mathcal{M}_{d\times d}$ denotes the Hessian matrix of f and diag$\mathsf{H}f$ denotes the diagonal matrix formed by replacing all off-diagonal entries of $\mathsf{H}f$ by zeros. Similarly to the univariate case, in order to (3.56) be valid, all entries in $\mathsf{H}f(\boldsymbol{x})$ must be square integrable. Also, if $n^{-1}|\mathbf{H}|^{-1/2} \to 0$ and $\mathbf{H} \to 0$ as $n \to \infty$ then MISE $\hat{f}(\cdot;\mathbf{H}) \to 0$.

Similarly to the univariate case, the critical step here is to estimate $\boldsymbol{\Psi}_4$, since this is the only unknown value in (3.56). Unfortunately, unlike the univariate case, it is not possible to derive a closed explicit formula for $\mathbf{H}_{\text{AMISE}}$ using (3.56) (an analogue for (3.45)). Historically, the first attempt at solving this problem was based on an observation that every element in the matrix $\boldsymbol{\Psi}_4$ can be estimated individually (the so-called *element-wise estimation* of $\boldsymbol{\Psi}_4$), see [187]. The drawback of this approach is that $\mathbf{H}$ is restricted only to the case of diagonal bandwidth matrices, with an exception of the bivariate case, where a closed solution was proposed. Another drawback, reported in [52], is that if the elements of the matrix $\boldsymbol{\Psi}_4$ are estimated separately, the resulting matrix estimate may not be positive definite (resulting in AMISE estimate not having a finite global minimum).

3.5 Adaptive Kernel Density Estimation

The basic definition of KDE (3.5) assumes that the bandwidth h is constant for every individual kernel (for illustrative example see Fig. 3.2). Small values of h result in slimming the kernels, while big values of h result in stretching them out. A useful extension is to use a different h depending on the local density of the input data points. This concept is known as *adaptive* or *variable* KDE. In the denser regions the bandwidth could be slimmed, while in the sparser regions it could be stretched out. The former allows one to emphasize the nature of KDE and the latter allows one to eliminate tails that are too fat.

Adaptive KDE can be grouped into two categories: *balloon* estimators, introduced in [113], and *sample point* estimators, first introduced in [14].

In the balloon estimator, a fixed bandwidth is selected in order to estimate f at a point x. Next, the estimate of f at x is calculated as an *average of identically scaled kernels* centered at each data point X_i. To calculate the estimate at another point x, a new value of the bandwidths is chosen to scale the kernels. This estimator is described in the following way

$$\hat{f}_B(x; h(x)) = \frac{1}{nh(x)} \sum_{i=1}^{n} K\left(\frac{x - X_i}{h(x)}\right). \tag{3.60}$$

Unfortunately, the balloon estimator suffers from a number of drawbacks the biggest one being that this estimator does not, in general, integrate to one over the entire domain. If the global PDF is needed, this can be a very serious problem. The literature review reveals a scarcity of sources devoted to the problem of choosing a practically acceptable $h(x)$. The MSE criterion (see [188, Sect. 2.10.1]), means that the asymptotically optimal bandwidth is

$$h_{\text{AMSE}}(x) = \left(\frac{R(K) f(x)}{\mu_2(K)^2 f''(x)^2 n}\right)^{1/5}, \tag{3.61}$$

provided $f''(x) \neq 0$. If, however, $f''(x) = 0$, then there are authors ([153, 188]) suggesting that one should take into account subsequent terms in the asymptotic expansion of MSE. On the other hand, [120] suggests that even with this correction, the optimal bandwidth is not defined at some points and goes on to propose a solution for this problem. It seems that the univariate KDE balloon estimators are not an interesting alternative for other KDE estimators (sample points and the classical one where the bandwidth h is fixed). The literature on the multivariate case is sparse, making it difficult to judge the practical usability of the multivariate balloon estimators.

Figure 3.13 demonstrates how the balloon KDE works. The five data points are

$$X_1 = -1.5, \; X_2 = -1, \; X_3 = -0.5, \; X_4 = 1, \; X_5 = 1.5, \tag{3.62}$$

and an arbitrary chosen bandwidth function is

$$h(x) = 0.5 + 1/(x^2 + 1). \tag{3.63}$$

The top left plot shows the $h(x)$ function. The top right plot shows the balloon KDE $\hat{f}_B(x; h(x))$. The last four plots show the kernels centered at each data point X_i and the KDE estimates at points $x = -1$, $x = 0$, $x = 0.5$ and $x = 2.5$. For every point x, a fixed bandwidth is chosen according to the $h(x)$ function (3.63).

Since the extension to the multivariate case is rather obvious, the details are omitted here.

The sample point estimator uses a different bandwidth for each data point X_i. The estimate of f at every x is then an *average of differently scaled kernels* centered at each data point X_i. This estimator is described in the following way

$$\hat{f}_{SP}(x; h(X_i)) = \frac{1}{n}\sum_{i=1}^{n}\frac{1}{h(X_i)}K\left(\frac{x - X_i}{h(X_i)}\right). \tag{3.64}$$

The sources [11, 99, 148, 178, 190] describe details of a more advanced research on the above, but only for univariate case. The multivariate case is studied for example in [149].

Sample points estimators are 'true' densities but can suffer from another drawback, that is the estimate at a certain point can be strongly affected by data located far from the estimation point (see [98]). However, this seems not to be a very serious problem in terms of practical applications and sample points estimators prove to be very useful (see simulation example below).

We now describe one of the possible approaches in terms of the sample point estimator (see [107, 151]):

1. Calculate $\hat{f}(x, h)$ in the usual way.
2. Determine a modification parameter s_i, $i = 1, 2, \ldots, n$, as

$$s_i = \left(\frac{\hat{f}(X_i, h)}{T}\right)^{-c}, \tag{3.65}$$

where $c \geq 0$ and T is the geometric mean of the values of the KDE of all data points, that is $\hat{f}(X_1, h), \hat{f}(X_2, h), \ldots, \hat{f}(X_n, h)$

$$T = \exp\left(\frac{1}{n}\sum_{i=1}^{n}\ln(\hat{f}(X_i, h))\right). \tag{3.66}$$

3. Finally, the sample point KDE is defined in the following way

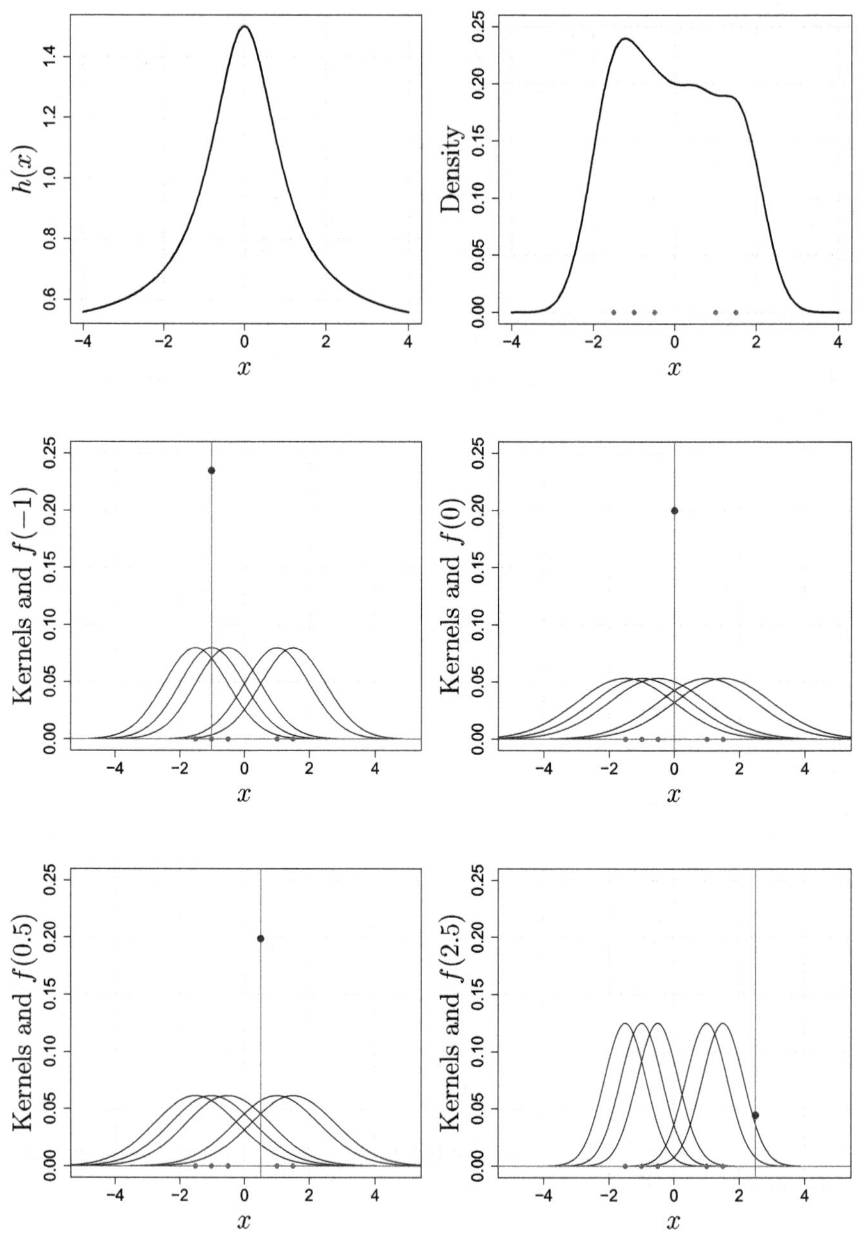

Fig. 3.13 A demonstration of the balloon KDE for a toy five points example

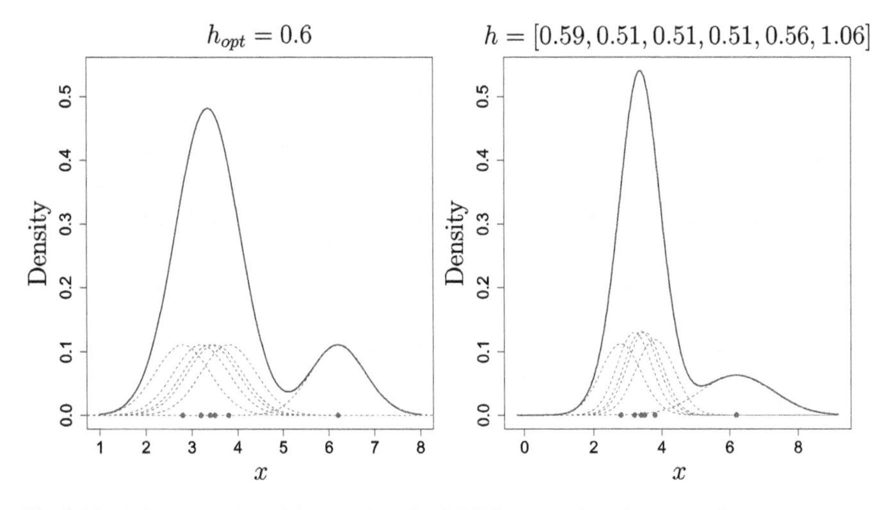

Fig. 3.14 A demonstration of the sample point KDE for a toy six points example

$$\hat{f}(x, h) = \frac{1}{nh} \sum_{i=1}^{n} \frac{1}{s_i} K\left(\frac{x - X_i}{hs_i}\right). \tag{3.67}$$

It is easy to notice that if $c = 0$, then $s_i \equiv 1$ and (3.5), (3.67) are the same. From (3.65) it is obvious that $s_i > 1$, if $\hat{f}(X_i, h)$ is smaller than the geometric mean of $\hat{f}(\cdot, h)$ resulting in the kernels being stretched out. In the opposite case, $s_i < 1$ and the kernels are slimmed. Finally, if $\hat{f}(X_i, h) = T$, then $s_i = 1$. In practice, the parameter c can be initially set to 0.5. Figure 3.14 demonstrates the sample point KDE. The six data points are

$$X_1 = 2.8, X_2 = 3.2, X_3 = 3.4, X_4 = 3.5, X_5 = 3.8, X_1 = 6.2, \tag{3.68}$$

and the nominal (optimal) bandwidth is $h_{opt} = 0.6$. It can be easily seen that the rightmost (the sixth) data point (it can be interpreted as an outlier) causes a disproportionately high local density value. By slightly stretching out its corresponding kernel bandwidth ($h_6 = 1.06 > h_{opt}$), the final density value in this point is reduced. On the other hand, the remaining bandwidths are slightly reduced and this results in generating more density mass in the region with more data points concentration.

Consider another example. Figure 3.15 shows the KDE of the log-normal distribution given by the following equation

$$f(x) = \mathcal{N}(\ln x; \mu = 0, \sigma = 1), \tag{3.69}$$

with $n = 200$ and $h = 0.3$. It can be easily seen that the classical KDE given by (3.5) contains a few spurious bums (left plot). Using the sample point KDE one eliminates this problem (right plot). The true density is plotted in the dashed line.

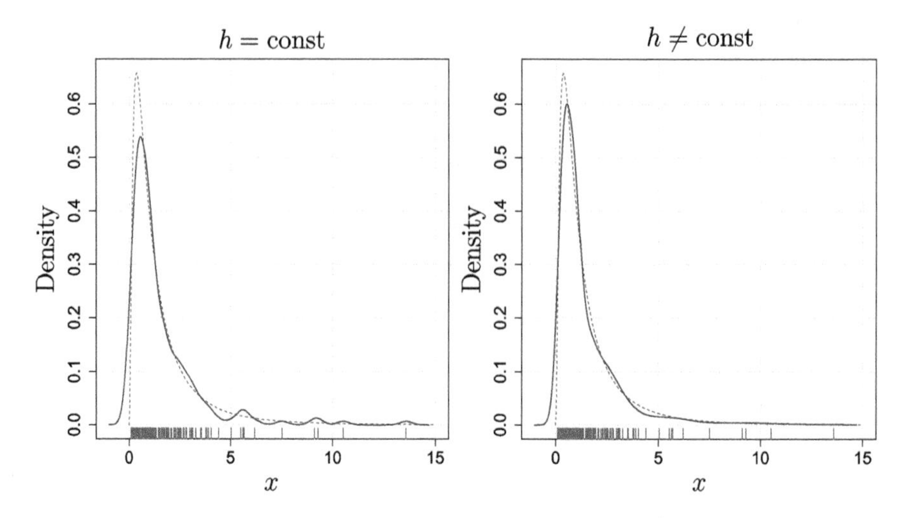

Fig. 3.15 A demonstration of the sample point KDE for the density $\mathcal{N}(\ln x; \mu = 0, \sigma = 1)$ with $n = 100$ and $h = 0.3$. The true density is plotted in the dashed line

3.6 Kernel Density Estimation with Boundary Correction

A general problem with KDE is that certain difficulties can arise at the boundaries and near them. In many practical situations the values of a random variable X are bounded. For example, the age of a person obviously cannot be a negative number. Likewise, duration of a technical process is always a positive number. On the other hand, the consecutive KDE will always have unlimited support, if the normal kernel is used where its domain ranges from $-\infty$ to $+\infty$. Even if a kernel with finite support is used (see Fig. 3.1), the consecutive KDE can usually go beyond the permissible domain. In such situations a naive approach of truncating $\hat{f}(x)$ to $[0, \infty]$ and setting $\hat{f}(x) = 0$ for all $x < 0$ is not acceptable as the resulting density does not integrate to one. A kind of scaling the consecutive KDE is also a bad idea as it is not obvious how such a scaling should look.

Below, after [100], we present a smart procedure based on 'reflection' of same unnecessary KDE parts. This idea is illustrated in Fig. 3.16. Let the admissible domain be $X \in [X_*, \infty]$. The kernel K plotted in the thin solid line refers to a data point X_i. Note that a part of this kernel lies outside the admissible domain and, as a consequence, the left-cut fragment of the kernel K does not integrate to one (the area $P1 + P2 < 1$). Therefore, the idea is to add the missing 'probability mass' represented by the area $P2$ to the kernel K. The reflected kernel K_r, plotted in the dashed line in the leftmost part of the figure, is centered at the point $2X_* - X_i$. Now, it is easy to notice that $P2 = P3$. Finally, the new kernel represented by the areas $P1 + P2 + P3$ integrates to one. From the point of view of analysis, it can be said that the above described procedure sums two kernels jointly with the indicator function. Equation (3.70) defines the left-side boundary correction. The right-side one (3.71)

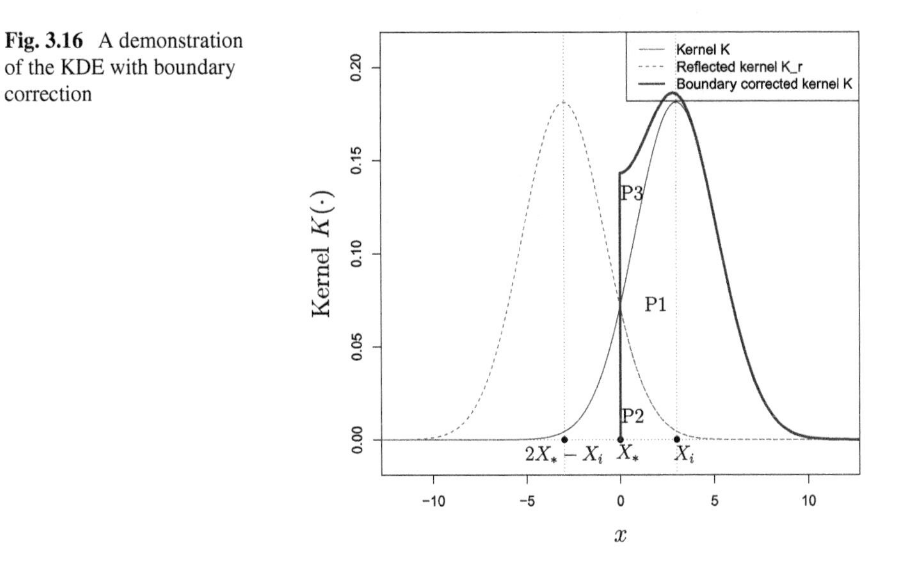

Fig. 3.16 A demonstration of the KDE with boundary correction

can be defined in a similar fashion (the only difference being that the different subset is used in the indicator function $\mathbb{1}(\cdot)$).

$$\hat{f}(x,h) = \frac{1}{nh}\sum_{i=1}^{n}\left[K\left(\frac{x-X_i}{h}\right) + K\left(\frac{x-(2X_*-X_i)}{h}\right)\right]\mathbb{1}_{\{x\in[X_*,\infty)\}}. \quad (3.70)$$

$$\hat{f}(x,h) = \frac{1}{nh}\sum_{i=1}^{n}\left[K\left(\frac{x-X_i}{h}\right) + K\left(\frac{x-(2X_*-X_i)}{h}\right)\right]\mathbb{1}_{\{x\in(-\infty,X_*]\}}. \quad (3.71)$$

Obviously, the adaptive KDE as described in Sect. 3.5 can also be used here. In such a case, the modified formulas are as follows

$$\hat{f}(x,h) = \frac{1}{nh}\sum_{i=1}^{n}\frac{1}{s_i}\left[K\left(\frac{x-X_i}{hs_i}\right) + K\left(\frac{x-(2X_*-X_i)}{hs_i}\right)\right]\mathbb{1}_{\{x\in[X_*,\infty)\}}, \quad (3.72)$$

and

$$\hat{f}(x,h) = \frac{1}{nh}\sum_{i=1}^{n}\frac{1}{s_i}\left[K\left(\frac{x-X_i}{hs_i}\right) + K\left(\frac{x-(2X_*-X_i)}{hs_i}\right)\right]\mathbb{1}_{\{x\in(-\infty,X_*]\}}. \quad (3.73)$$

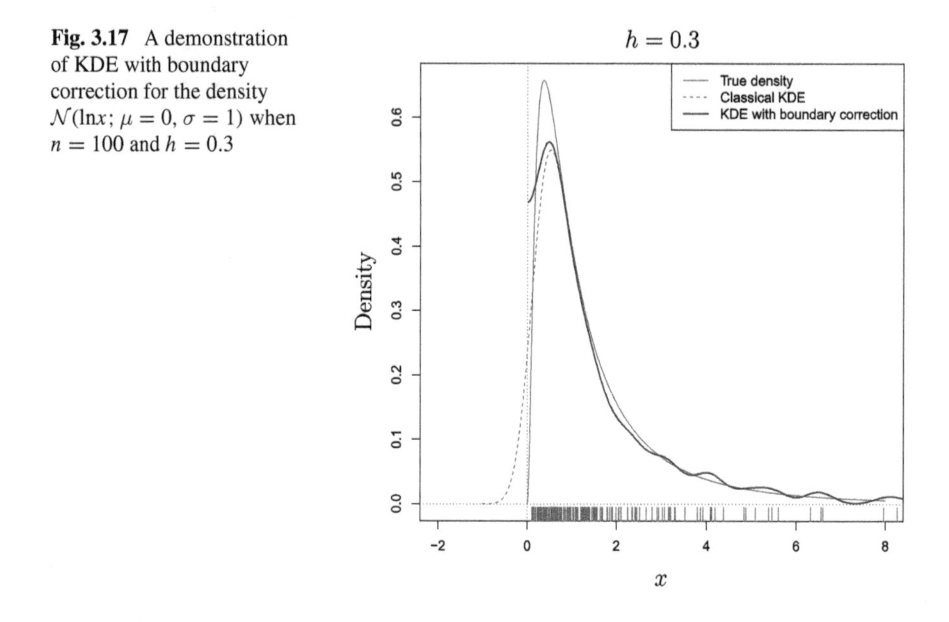

Fig. 3.17 A demonstration of KDE with boundary correction for the density $\mathcal{N}(\ln x; \mu = 0, \sigma = 1)$ when $n = 100$ and $h = 0.3$

Figure 3.17 shows the KDE of the log-normal distribution given by (3.69) with $n = 200$ and $h = 0.3$. It can be easily seen that the classical KDE given by (3.5) includes (incorrectly) the inaccessible domain $(-\infty, 0]$. Using KDE with boundary correction (left-side) eliminates this problem. The problem of boundary effects has been studied by many authors, see for example [33, 79, 92, 100, 107, 116, 122, 197–199].

Extending the above to the multivariate case is rather obvious, therefore the details of this procedure are omitted here.

3.7 Kernel Cumulative Distribution Function Estimation

The techniques used in the context of KDE can be easily extended to take into account various other probability distribution functions. In this section, we describe an extension to the estimation of the *cumulative distribution function* (CDF) (see [107]). The CDF of a real-valued random variable X evaluated at x is defined as the probability of X taking a value less than or equal to x, that is

$$F(x) = P(X \le x) = \int_{-\infty}^{x} f(u)du. \tag{3.74}$$

Every CDF is a non-decreasing and right-continuous function, additionally

$$\lim_{x \to -\infty} F(x) = 0,$$

$$\lim_{x \to +\infty} F(x) = 1. \tag{3.75}$$

Therefore, the univariate kernel CDF estimator (KCDE) for a random sample $X_1, X_2, \ldots X_n$ drawn from a common and usually unknown density f can be given by

$$\hat{F}(x) = \int_{-\infty}^{x} \hat{f}(u) du, \tag{3.76}$$

and, after substituting to (3.76), the equation for KDE given by (3.5), the following KCDE is obtained

$$\hat{F}(x) = \frac{1}{nh} \sum_{i=1}^{n} \int_{-\infty}^{x} K\left(\frac{u - X_i}{h}\right) du. \tag{3.77}$$

Let I be an *antiderivative* of the kernel function K, that is

$$I(x) = \int_{-\infty}^{x} K(u) du, \tag{3.78}$$

and finally we arrive at the KCDE, given by

$$\hat{F}(x, h) = \frac{1}{nh} \sum_{i=1}^{n} I\left(\frac{x - X_i}{h}\right). \tag{3.79}$$

Unlike for KDE, in this case the Gaussian function cannot be considered for the kernel K, since the antiderivative function cannot be expressed in terms of elementary functions (only numerical approximations are possible). Fortunately, it is easy to derive the antiderivative function in case of other commonly used kernel functions. For example, in case of the Epanechnikov kernel it is not difficult to show that

$$I(x) = \begin{cases} 0 & \text{for } x \in (-\infty, -1) \\ 0.25(-x^3 + 3x + 2) & \text{for } x \in [-1, 1] \\ 1 & \text{for } x \in (1, \infty). \end{cases} \tag{3.80}$$

As an example of how KCDE works, consider a toy univariate dataset (3.11). The KCDE based on these data is depicted in Fig. 3.18. The bandwidth h was calculated using the univariate plug-in rule implemented in the *ks R* package yielding $h = 1.24$, see [3, 48, 49]. The antiderivative functions I are drawn in the dashed lines, the data points are marked by small dots on the x axis.

Extending this description to cover the multivariate case is rather obvious and therefore we skip the details of this procedure. Moreover, the adaptive approach and

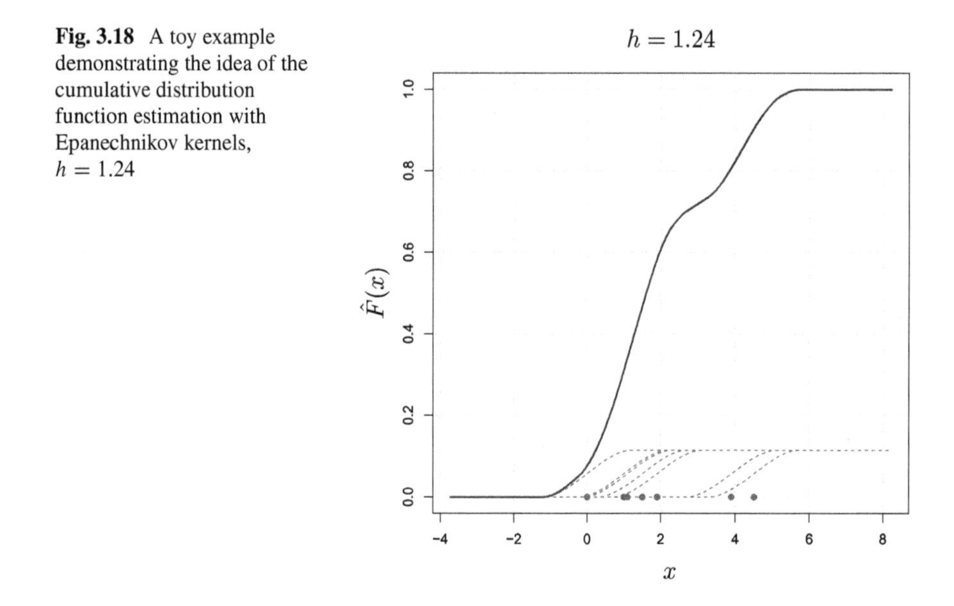

Fig. 3.18 A toy example demonstrating the idea of the cumulative distribution function estimation with Epanechnikov kernels, $h = 1.24$

the boundary correction approach can also by used here, as presented in Sects. 3.5 and 3.6, respectively.

3.8 Kernel Density Derivative Estimation

The *kernel density derivative estimation* (KDDE) (see [9, 101, 154]) is a natural extension of KDE as defined by (3.5). If the kernel K is differentiable r times, then a natural estimator of the r-th derivative of $f(x)$ is

$$\hat{f}^{(r)}(x, h) = \frac{1}{nh^{r+1}} \sum_{i=1}^{n} K^{(r)} \left(\frac{x - X_i}{h} \right). \tag{3.81}$$

This estimator only makes sense, if $K^{(r)}(u)$ exists and is non-zero. Since the Gaussian kernel has derivatives of all orders, this is a common choice for derivative estimation. A multivariate extension, comprehensively analyzed in [21, 23, 50], is defined as follows

$$\begin{aligned} \mathsf{D}^{\otimes r} \hat{f}(\boldsymbol{x}, \mathbf{H}) &= n^{-1} \sum_{i=1}^{n} \mathsf{D}^{\otimes r} K_{\mathbf{H}} (\boldsymbol{x} - \boldsymbol{X}_i) \\ &= n^{-1} (\mathbf{H}^{-1/2})^{\otimes r} \sum_{i=1}^{n} (\mathsf{D}^{\otimes r} K)_{\mathbf{H}} (\boldsymbol{x} - \boldsymbol{X}_i) \end{aligned}$$

$$= n^{-1}|\mathbf{H}|^{-1/2}(\mathbf{H}^{-1/2})^{\otimes r} \sum_{i=1}^{n} \mathsf{D}^{\otimes r} K(\mathbf{H}^{-1/2}(\boldsymbol{x} - \boldsymbol{X}_i)). \tag{3.82}$$

Here, D is the gradient operator, the notation using the $\otimes$ symbol is explained in details in Sect. 4.3.3 where the above-mentioned generalization is presented.

The estimation of density derivatives is a very important problem and has a number of applications. First, the derivatives of the density function provide, in their own right, supplementary information about the data. The estimation of the gradient vector (the first derivative) can be used for data clustering and/or filtering, and it seems to be its most popular use (see examples in Sect. 7.3). Another very interesting problem where KDDE is used is the analysis of flow cytometry data. We give a brief overview of this problem in Sect. 7.6.

Similarly to the zero-th order KDE, the problem of an optimal bandwidth selection is a crucial factor for the KDDE. The main publication dealing with this is [21], where the unconstrained bandwidth selectors for the KDDE have been established.

Note that we omit many of the topics related to the KDDE, the details can be found in the quoted literature. The following [23, 83, 101] can be considered a good starting point.

3.9 The Curse of Dimensionality

The term *curse of dimensionality* was coined by Bellman [8] and refers to various phenomena and problems that arise when analyzing data in high-dimensional spaces that are usually absent in low-dimensional spaces (say, below four). In the context of KDE, the curse of dimensionality manifests itself in the following way: an enormous amount of data is required to learn plausible probability density functions. In high dimensions data are extremely sparse and distance measure becomes meaningless.

The problem with data sparseness can be, at least theoretically, neutralized by increasing dataset sizes (for example collecting more measurements, if possible). It is also possible to assess a minimal sample size to achieve a given level of accuracy [170, p. 125], [168, p. 94] and [157, pp. 201–202]. The calculations presented in these publications assume that the unit multivariate normal density functions are used and that the kernel is normal. In this sense, the minimal sample sizes obtained are very optimistic and for real datasets should certainly be magnified. The results described in these publications differ slightly from each other (depending on a methodology used) but in either case the minimal sample sizes increase rapidly with dimensionality d.

Next problematic feature in high dimensions is a counter-intuitive *data concentration in a hypercube*. To explain this phenomena let us suppose that a d-dimensional *hypercube* is given with edges of length $2r$. The hypercube has 2^d corners and its volume is

$$V_{\text{cube}}(2r) = (2r)^d. \tag{3.83}$$

Now, consider the *biggest hypersphere* with radius r (in two and three dimensions these are a circle and a ball, respectively) that can be inscribed in the hypercube. The volume of such a sphere is

$$V_{\text{sphere}}(r) = \frac{r^d \pi^{d/2}}{\Gamma(1 + d/2)}, \tag{3.84}$$

where Γ is the Gamma function [1] defined as

$$\Gamma(z) = \int_0^\infty x^{z-1} e^{-x}\, dx. \tag{3.85}$$

Figure 3.19 shows the volume of the unit hypersphere (3.84) with increasing dimensionality. At first glance, the result is somewhat unexpected and counter-intuitive. Initially, the volume increases, then achieves the highest value for $d = 5$ and finally the volume decreases rapidly approaching zero for $d \to \infty$. It is also true that as the dimensionality d increases asymptotically

$$\lim_{d\to\infty} \frac{V_{\text{sphere}}(r)}{V_{\text{cube}}(2r)} = \lim_{d\to\infty} \frac{r^d\ \pi^{d/2}}{2^d\ \Gamma(1 + d/2)} \to 0. \tag{3.86}$$

This means that as the dimensionality increases, most of the volume of the hypercube is located very near the corners and the center is almost empty. In other words, most of the probability mass is concentrated in the tails of PDF. For a nice conceptual overview of high-dimensional spaces see [196, Chap. 6]. For example, in case of a 10-dimensional normal distribution $\mathcal{N}(\mathbf{0}_{10}, \mathbf{1}_{10})$ only about 2% of the total probability mass lies within a band around the mean with a width of two standard deviations (for comparison, in case of the one-dimensional normal distribution this is about 68%).

Another feature of high-dimensional spaces is that the distance measure becomes meaningless and typical distance measures (e.g. Euclidean distance, Mahalanobis distance, Manhattan distance etc.) loose their validity as metrics. Formally, the above can be written as

$$\lim_{d\to\infty} \frac{\text{dist}_{\max}(d) - \text{dist}_{\min}(d)}{\text{dist}_{\min}(d)} \to 0, \tag{3.87}$$

where $\text{dist}_{\max}(d)$ and $\text{dist}_{\min}(d)$ are the minimum and the maximum distances between a reference point and all other points in a dataset. Therefore, in spaces of high dimensions, the distance measures start loosing their effectiveness in measuring dissimilarity.

One concludes that the nonparametric methods for kernel density problems should not be used for high-dimensional data and it seems that a feasible dimensionality should not exceed five or six (this view is in line with the views expressed by many researchers focusing on nonparametric methods).

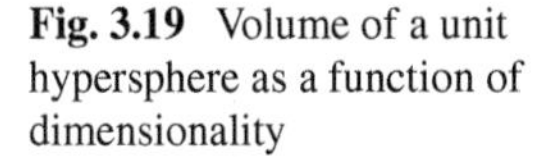

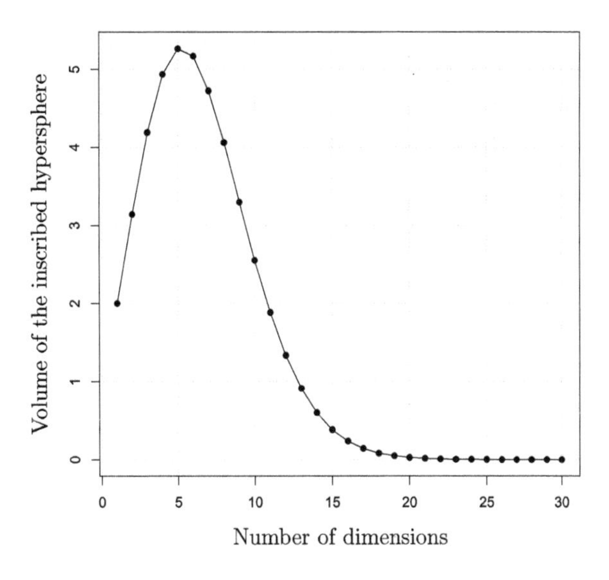

Fig. 3.19 Volume of a unit hypersphere as a function of dimensionality

3.10 Computational Aspects

It is obvious from (3.5) and (3.15) that the naive direct evaluation of KDE at m evaluation points for n data points requires $O(mn)$ kernel evaluations and $O(mn)$ additions and multiplications. Evaluation points can be of course the same as data points and then the resulting computational complexity can be calculated to be $O(n^2)$, making it a very expensive trade-off, especially for large datasets and higher dimensions. The same relates to the case of the kernel density derivative estimation (3.82).

A typical approach would be to calculate KDE for *equally spaced grid* of evaluation points. This approximation technique is known as *binning* and is described in details in Sect. 5.2. Binning is the most convenient way in case a bivariate or trivariate contour or perspective plots are needed. To obtain the final plots providing a reasonable approximation, the number of grid points should be not less that $g = 50$ in every dimension, giving g^d individual evaluation points. In such cases, even for small or moderate sample sizes (tens or hundreds), the computational complexity grows rapidly and the naive evaluation of KDE is totally impractical. Of course, for large datasets, when $g^d \ll n$, the binning provides significant computational advantages. A combination of two techniques, namely fast Fourier transform (FFT) and binning can dramatically increase the computational speed of computing KDE [72] (see also Chap. 5).

There has been little research devoted to the fast KDE computation as compared to other similar topics, see for example [144] for a comprehensive overview. An attempt at using the *Message Passing Interface* (MPI) was presented in [114]. In [144], the authors give an ϵ-*exact* approximation algorithm, where the constant ϵ controls the desired arbitrary accuracy. Other techniques, such as those proposing the use of *Graphics Processing Units* (GPUs) [5] and *Field-Programmable Gate Arrays* (FPGA) [73] have also been taken into account. Chapter 6 describes the FPGA-based results in more detail.

Chapter 4
Bandwidth Selectors for Kernel Density Estimation

4.1 Introduction

The problem of choosing the bandwidth parameter is a crucial issue that occurs often in the context of KDE. In the previous chapters, the assumption that this parameter is known has been made. This chapter presents an overview of the existing methods for optimal bandwidth selection. The subject literature is abundant in many different solutions. Unfortunately, there does not exist one best method that can be applied universally.

The accuracy of KDE depends very strongly on the bandwidth value. In the univariate case, the bandwidth is a *scalar* that controls the amount of smoothing. In the multivariate case, the bandwidth is a *matrix* and it controls both the amount and the orientation of smoothing. This matrix can be defined on various levels of complexity. The simplest case is when a positive constant scalarbandwidth!positive constant scalar multiplies the identity matrix, that is, $\mathbf{H} \in \mathcal{S}$, where $\mathcal{S} = \{h^2\mathbf{I_d} : h > 0\}$. Another level of sophistication is to consider $\mathbf{H} \in \mathcal{D}$. These two forms are often called *constrained*. In the most general form, the bandwidth is *unconstrained*, that is $\mathbf{H} \in \mathcal{F}$ (see also Sect. 3.3.2). A very important problem to be considered before evaluating (3.5) or (3.15) is to find the optimal (given certain criteria) bandwidth. There are many ways of doing so: most of them can be described as *data-driven bandwidth selectors*.

The problem of selecting the scalar bandwidth for the univariate kernel density estimation is thoroughly analyzed. A number of methods exist that combine good theoretical properties with strong practical performance. See for example [15, 103, 104, 181, 188], where one can find a comprehensive history of the development of such selectors. Many of these univariate selectors can be extended to the multivariate case in a relatively straightforward fashion, if $\mathbf{H}$ is constrained (see [150, 187]). However, if $\mathbf{H}$ is unconstrained, such generalization is not that easy. Comprehensive analysis of unconstrained bandwidth selectors was made mainly in [19–22, 46, 53, 54]. These papers reference all main bandwidth selectors, hence we do not

A. Gramacki, *Nonparametric Kernel Density Estimation and Its Computational Aspects*, Studies in Big Data 37,
https://doi.org/10.1007/978-3-319-71688-6_4

describe them here and the reader is invited to consult the above for more information. A comprehensive review and comparison of fully automatic bandwidth selectors can be found in [89] (only the univariate case is described).

The three major types of bandwidth selectors are as follows:

1. Methods using very simple and easy to compute mathematical formulas. They were developed to cover a wide range of situations, but do not guarantee that the result is close enough to the optimal (under certain criteria) bandwidth. They are often called the *rules-of-thumb* (ROT), see [158, 168].
2. Methods based on the notion of *cross-validation* (CV) that are based on a more precise mathematical footing (see [12, 147]). They require much more computational power, providing, however, the bandwidths that are more accurate for a wider class of density functions. Three classical variants of the CV are: *least squares cross validation* (LSCV), sometimes called *unbiased cross validation* (UCV), see [162], *biased cross validation* (BCV), see [162], and *smoothed cross validation* (SCV), see [77].
3. Methods based on plugging in estimates of some unknown quantities that appear in formulas for the asymptotically optimal bandwidth. They are often called *plug-in* (PI), see [129, 165].

Almost all modern bandwidth selectors share the basic idea of using the AMISE error criteria as a starting point (for the univariate case (3.44), (3.47) and for the multivariate case (3.56)). The critical step is to estimate Ψ_4 or $\boldsymbol{\Psi}_4$ as these are the only unknown values in the equations.

The chapter is organized as follows: Sect. 4.2 presents the most popular univariate bandwidth selectors. Section 4.3 is devoted to the multivariate bandwidth selectors. Section 4.4 discusses some computational aspects of the bandwidth selection task.

4.2 Univariate Bandwidth Selectors

The problem of the bandwidth selection in the univariate case is a well-researched area having been studied for many years. There is a number of now-classical methods that are commonly used by in practice giving very good results. In Sect. 4.1 the three classes of bandwidth selectors were mentioned (ROT, CV and PI). Below, we provide their brief description.

4.2.1 Univariate Rule-of-Thumb Selectors

The results obtained using (3.45) depend on an unknown density f. The *rule-of-thumb* selector (also known as *normal scale* selector) replaces the unknown density function f by a reference distribution function. In other words, an unknown parameter is replaced with its estimate. Typically, the assumption that the unknown density f belongs to the family of normal distribution with mean μ and variance σ^2 is made. After certain manipulations, for details see [84] and [188], it can be shown that

$$\begin{aligned} h_{\text{AMISE}} = \hat{h}_{\text{ROT1}} &= \left(\frac{8\pi^{1/2} R(K)}{3\mu_2(K)^2} \right)^{1/5} \hat{\sigma}\, n^{-1/5} \\ &\approx 1.06\, \hat{\sigma}\, n^{-1/5}, \end{aligned} \tag{4.1}$$

where $\hat{\sigma}$ is the estimate of σ (the standard deviation of the input data). While this rule-of-thumb is easy to compute, it should be used with caution, as it can yield widely inaccurate estimates when the density is not close to being normal. What is achieved by working under the normality assumption is an explicit and very easy to use formula for bandwidth selection. In practice, the true distribution of X is usually unknown and in such a case (4.1) gives a bandwidth close to the optimal one only if the distribution of X is not significantly different from the normal distribution.

A problem may occur in case when a dataset has outliers, as (4.1) is very sensitive to their presence. The problem is related to the need to estimate σ. Any outlier can affect the value of $\hat{\sigma}$ to be too large and, consequently, producing a bandwidth that is too large as well. A version that is not that affected by the presence of outliers can be proposed, in cases, where the interquartile range $\text{IQR} = Q_3 - Q_1$ is used as a measure of the data spread instead of the variance [168, Sect. 3.4.2]. Hence, the improved version of the rule-of-thumb is

$$h_{\text{AMISE}} = \hat{h}_{\text{ROT2}} = 1.06 \min\left\{ \hat{\sigma}, \frac{\text{IQR}}{1.34} \right\} n^{-1/5}. \tag{4.2}$$

4.2.2 *Univariate Plug-In Selectors*

Plug-in selectors perform extremely well. They also enjoy a number of desirable properties, both in terms of theory and practice, especially in the context of a fast rate of convergence and a low sampling variability (see Fig. 4.2 for an illustrative example). This makes these methods a common first choice in practical applications. These classes of selectors are based on the AMISE criterion (3.47). In this equation, the only unknown is Ψ_4: it must be estimated first, and this is the critical step of all plug-in selectors. The most commonly used method of estimating Ψ_4 was introduced in [165]. According to this method, the estimator $\hat{\Psi}_4$ ($\hat{\Psi}_r$ in general, where r is a derivative order) is

$$\hat{\Psi}_r(g_r) = n^{-1} \sum_{i=1}^{n} \hat{f}^{(r)}(X_i; g_r) = n^{-2} \sum_{i=1}^{n} \sum_{j=1}^{n} L_{g_r}^{(r)}(X_i - X_j), \tag{4.3}$$

where g_r and L_{g_r} are *pilot bandwidth* and *pilot kernel*, respectively. In general, g_r and L_{g_r} can be different than h and K but in practical computer implementations these are almost always the same ([49, 191]). After replacing Ψ_4 in (3.47) by its estimator $\hat{\Psi}_4$ one obtains the *direct plug-in* (DPI) or *solve-the equation* (STE) estimators.

Direct plug-in selector (DPI)

The starting point of the DPI is

$$h_{\text{AMISE}} = \left(\frac{R(K)}{\mu_2(K)^2 \hat{\Psi}_4(g_4) n} \right)^{1/5}. \tag{4.4}$$

This equation cannot be used directly as it depends on g_4, the value of which is unknown. It was shown in [188] that g_r can be computed using the AMSE-optimal bandwidth formula (assuming that $L = K$)

$$g_{r,\text{AMSE}} = \left[\frac{2K^{(r)}(0)}{-\mu_2(K)\Psi_{r+2}(g_{r+2})n} \right]^{1/(r+3)}. \tag{4.5}$$

So,

$$g_{4,\text{AMSE}} = \left[\frac{2K^{(4)}(0)}{-\mu_2(K)\Psi_6(g_6)n} \right]^{1/7}, \tag{4.6}$$

and, similarly to the above, $\Psi_6(g_6)$ depends on the unknown value g_6. In that case, $g_{6,\text{AMSE}}$ can be computed as

$$g_{6,\text{AMSE}} = \left[\frac{2K^{(6)}(0)}{-\mu_2(K)\Psi_8(g_8)n} \right]^{1/9}. \tag{4.7}$$

The above sequence can be continued but the classical DPI stops at this point and assumes that $\Psi_8(g_8)$ is calculated according to the following *normal scale* formula (valid where r is an even number)

$$\Psi_r^{\text{NS}} = \left[\frac{-1^{r/2} r!}{(2\hat{\sigma})^{r+1}(r/2)!\pi^{1/2}} \right]^{1/9}, \tag{4.8}$$

where $\hat{\sigma}$ is the estimate of the standard deviation σ. Finally, the bandwidth can be computed from (4.4).

The above procedure is the 2-stage DPI ($l = 2$). Larger l is impractical, since as l increases, with the bandwidth becoming less biased, it also becomes more variable. There is no established analytical method for selecting the optimal l and typically l is set to $l = 2$, which ensures very good practical performance. However, in [24] the authors describe an automatic (i.e. data-driven) method for choosing the number of stages to be employed in the plug-in bandwidth selector.

In summary, the DPI procedure relies on the following sequence of operations:

Step 1: Estimate $\Psi_8(g_8)$ using the normal scale estimate $\hat{\Psi}_8^{\text{NS}}$ (formula (4.8)).

Step 2: Estimate $\Psi_6(g_6)$ using the kernel estimator $\hat{\Psi}_6(g_6)$, where g_6 can be calculated from (4.7).

Step 3: Estimate $\Psi_4(g_4)$ using the kernel estimator $\hat{\Psi}_4(g_4)$, where g_4 can be calculated from (4.6).

Step 4: The bandwidth $\hat{h}_{\text{DPI},2}$ can now be calculated from (4.4).

Solve-the-equation selector (STE)

The choice of the starting point for the STE is also based on the AMISE-optimal criterion but with an additional requirement that the pilot bandwidth for the estimation of Ψ_4 is a function γ of h, that is

$$h_{\text{AMISE}} = \left(\frac{R(K)}{\mu_2(K)^2 \hat{\Psi}_4(\gamma(h))n} \right)^{1/5}, \tag{4.9}$$

where

$$\gamma(h) = \left(\frac{2K^{(4)}(0)\mu_2(K)\hat{\Psi}_4(g_4)}{-\hat{\Psi}_6(g_6)R(K)} \right)^{1/7} h^{5/7}. \tag{4.10}$$

Here, $\hat{\Psi}_4(g_4)$ and $\hat{\Psi}_6(g_6)$ are the kernel estimates of Ψ_4 and Ψ_6, respectively. The choice of g_4 and d_6 can be made using (4.6) and (4.7). Considerations regarding the number of stages similar to that in case of the DPI algorithm lead to the following 2-stage STE ($l = 2$) bandwidth selector:

Step 1: Estimate $\Psi_4(g_6)$ and $\Psi_8(g_8)$ using the normal scale estimates $\hat{\Psi}_6^{\text{NS}}$ and $\hat{\Psi}_8^{\text{NS}}$ (formula (4.8)).

Step 2: Estimate $\Psi_4(g_4)$ and $\Psi_6(g_6)$ using the kernel estimators $\hat{\Psi}_4(g_4)$ and $\hat{\Psi}_6(g_6)$ where g_4 and g_6 can be calculated from (4.6) and (4.7), respectively.

Step 3: Estimate $\Psi_4(g_4)$ using the kernel estimator $\hat{\Psi}_4(\gamma(h))$ where $\gamma(h)$ can be calculated from (4.10).

Step 4: The bandwidth $\hat{h}_{\text{STE},2}$ is the (numerical) solution of the equation (4.9).

4.2.3 Univariate Least Squares Cross Validation Selector

The least squares cross validation selector was developed independently in [12, 147] and is based on the notion of cross validation. Since it is not biased, the LSCV selector is sometimes called the *unbiased cross validation* selector. The analysis starts from (3.34) that can be rewritten as (for the sake of simplicity of notation, the integral limits are omitted):

$$\text{ISE}\, \hat{f}(\cdot; h) = \int \hat{f}(x; h)^2\, dx - \int \hat{f}(x; h) f(x)\, dx + \int f(x)^2. \tag{4.11}$$

The last integral does not depend on h and can be ignored and the first integral can be calculated from the data. This means that it is only the middle integral that has to be estimated. It depends on h and involves the unknown quantity f. It can be shown that the so-called *least square cross-validation criterion* for finding an optimal h is

$$\text{LSCV}(h) = \int \hat{f}(x; h)^2\, dx - 2n^{-1} \sum_{i=1}^{n} \hat{f}_{-i}(X_i; h), \tag{4.12}$$

where

$$\hat{f}_{-i}(x; h) = h^{-1}(n-1)^{-1} \sum_{j=1, i \neq j}^{n} K\left(\frac{x - X_j}{h}\right). \tag{4.13}$$

Here, $\hat{f}_{-i}(x; h)$ is the so-called *leave-one-out* density estimator. As the name suggests, the X_i observation is not used in the calculations. The name *cross-validation* refers to the use of one subset of data to make analysis on another subset. This ensures that the observations used to calculate $\hat{f}_{-i}(x; h)$ are independent of X_i.

To derive a practically usable cross-validation criterion, one needs to replace $\int \hat{f}(x; h)^2 dx$ by a quantity that employs sums rather than the integral. It can be shown [82, p. 230] that

$$\int \hat{f}(x; h)^2\, dx = h^{-1} n^{-2} \sum_{i=1}^{n} \sum_{j=1}^{n} K * K\left(\frac{X_i - X_j}{h}\right), \tag{4.14}$$

where $K * K(u)$ is the convolution of K with itself, that is

$$K * K(x) = \int K(u) K(x - u)\, du. \tag{4.15}$$

For normal kernels $K(x) = \mathcal{N}(x; 0, \sigma^2)$ that often appear in practical applications, it is true that

$$K * K(x) = \mathcal{N}(x; 0, 2\sigma^2). \tag{4.16}$$

Putting it all together, one obtains the final cross-validation criterion (or *objective function*) for finding the optimal h

$$\mathrm{LSCV}(h) = \frac{1}{h}\Bigg[\frac{1}{n^2}\sum_{i=1}^{n}\sum_{j=1}^{n} K * K\left(\frac{X_i - X_j}{h}\right) - \frac{2}{n(n-1)}\sum_{i=1}^{n}\sum_{j=1, j\neq i}^{n} K\left(\frac{X_i - X_j}{h}\right)\Bigg]. \tag{4.17}$$

The LSCV bandwidth scalar $\hat{h}_{\mathrm{LSCV}}$ is the minimizer of the objective function $\mathrm{LSCV}(h)$, that is

$$\hat{h}_{\mathrm{LSCV}} = \underset{h \in \mathbb{R}^+}{\operatorname{argmin}}\, \mathrm{LSCV}(h). \tag{4.18}$$

A version of (4.17) which is sometimes more convenient for fast computer implementations (see Chap. 5) can be rewritten in a slightly different form. The goal is to remove the unwanted condition $j \neq i$ in the second double summation. For a sufficiently large n (around the value of several dozen or so in practical applications) it is safe to assume that $n \approx n - 1$. Under this assumption, the $\mathrm{LSCV}(h)$ objective function has the following form

$$\mathrm{LSCV}(h) = \frac{1}{h}\Bigg[\frac{1}{n^2}\sum_{i=1}^{n}\sum_{j=1}^{n} K * K\left(\frac{X_i - X_j}{h}\right) - \frac{2}{n^2}\sum_{i=1}^{n}\sum_{j=1}^{n} K\left(\frac{X_i - X_j}{h}\right) + \frac{2}{n}K(0)\Bigg]. \tag{4.19}$$

The LSCV method is very popular in practical applications, mainly due to the fact that it is considered quite intuitive. However, this variant of the CV selector has some serious drawbacks, which has been pointed out by many authors. The drawbacks include having more than one local minimum in the objective function and high variability (in the sense that for different datasets from the same distribution, it typically gives considerably different answers). This behavior is demonstrated in Fig. 4.1. It shows the $\mathrm{LSCV}(h)$ versus h defined by (4.17) for a couple of samples drawn from $\mathcal{N}(0, 1)$ of size $n = 100$ using the normal kernel. The dashed vertical lines show the position of h_{MISE} defined by (3.45). The solid vertical lines show the position of the global minimum of the $\mathrm{LSCV}(h)$ indicated by $\hat{h}_{\mathrm{LSCV}}$. It can be easily observed that in many situations the global minimum is far from h_{MISE}, see Figs. 4.1a, c, h, j. On the other hand, Fig. 4.1b suggests that the best option for $\hat{h}_{\mathrm{LSCV}}$ would be the *largest local minima* and Figs. 4.1c, h suggest that the *inflection point* could be the best option for $\hat{h}_{\mathrm{LSCV}}$. In any case, this simple experiment shows that $\hat{h}_{\mathrm{LSCV}}$ is highly variable and as a result the LSCV selector should be used with care.

Figure 4.2 shows the values of $h_{\mathrm{MISE}} - h$ for LSCV and PI (the DPI variant) methods and different n. The high variability of the former is obvious, especially for large datasets. It should be noted that the optimal bandwidth often generates undersmoothed density estimates [29, 43]. Yet another drawback of the LSCV is

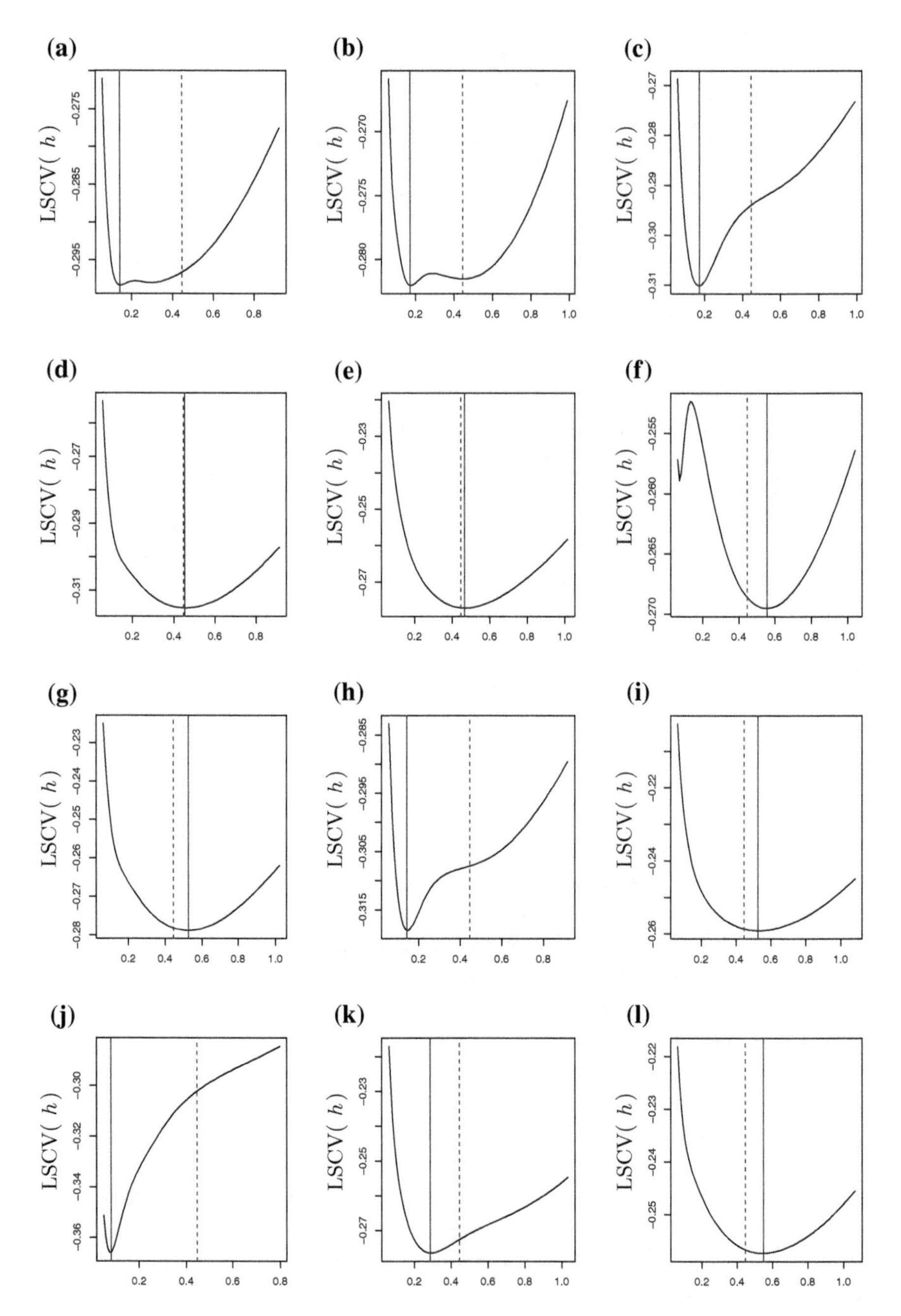

Fig. 4.1 LSCV(h) versus h for a number of samples drawn from $\mathcal{N}(0, 1)$ of size $n = 100$

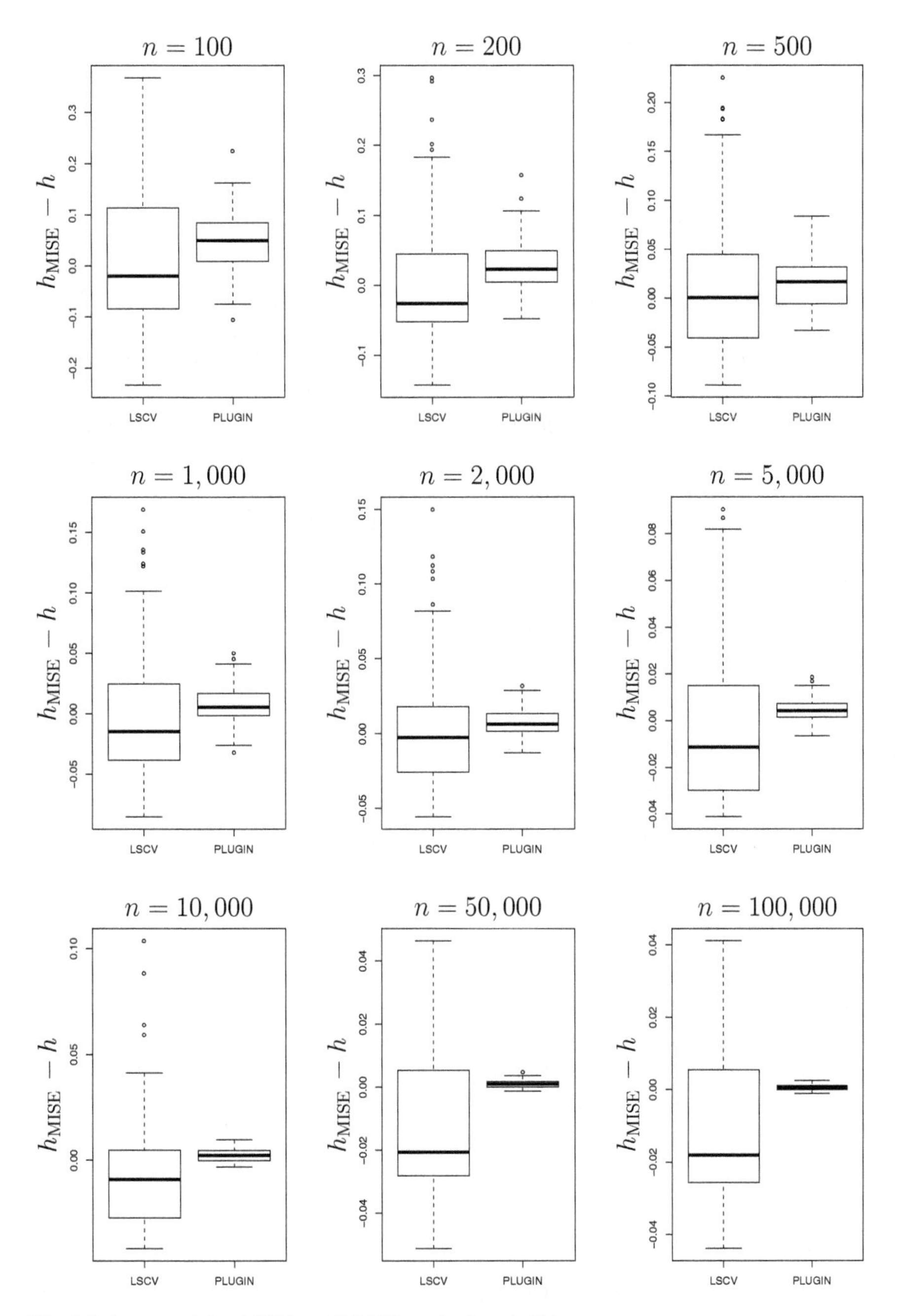

Fig. 4.2 $h_{\text{MISE}} - h$ for LSCV and PI(DPI) methods and different n

that it can give poor results if applied to discretized data (real data are nearly always rounded or discretized) when some replications are very possible. This phenomenon is explained in [168, pp. 51–52].

4.2.4 Univariate Smoothed Cross Validation Selector

Smoothed cross validation selector (SCV) is a version of the general cross-validation selector. Its 'smoothness' means that compared to its 'not-smoothed' counterpart selector (LSCV/UCV), the variability of the resulted KDE is reduced. SCV is a somehow similar to the PI selector in the sense that it uses a kernel estimator with pilot bandwidth g to estimate the integrated squared bias component of the MISE $\hat{f}(\cdot; h)$. The difference is that the SCV is based on the exact integrated squared bias rather than its asymptotic approximation. It also comprises the asymptotic integrated variance $(nh)^{-1}R(K)$, so the objective function to be minimized is

$$\mathrm{SCV}(h) = (nh)^{-1}R(K) + \mathrm{ISB}(h), \tag{4.20}$$

where

$$\mathrm{ISB}(h) = \int (K_h * \hat{f}_L(\cdot; g) - \hat{f}_L(\cdot; g))^2 \, dx, \tag{4.21}$$

is the estimate of the integrated squared bias (ISB). It can be proved that

$$\begin{aligned} ISB(h) = n^{-2} \sum_{i=1}^{n} \sum_{j=1}^{n} (K_h * K_h * L_g * L_g \\ - 2 * K_h * L_g * L_g + L_g * L_g)(X_i - X_j), \end{aligned} \tag{4.22}$$

where g and L are the *pilot bandwidth* and the *pilot kernel*, respectively. In general, g and L can be different than h and K but in practical computer implementations these are almost always the same. In that case, if $L = K$ and K is the normal kernel, i.e. $K = \phi$ and using the properties of the convolution of two univariate Gaussians, that is

$$\begin{aligned} &\text{if} \quad f(x) = \mathcal{N}(\mu_f, \sigma_f^2) \\ &\text{and} \quad g(x) = \mathcal{N}(\mu_g, \sigma_g^2) \\ &\text{than} \quad f(x) * g(x) = \mathcal{N}\left(\mu_f + \mu_g, \sqrt{\sigma_f^2 + \sigma_g^2}\right), \end{aligned} \tag{4.23}$$

the objective function obtains the following form

$$\text{SCV}(h) = (nh)^{-1}(2\pi^{1/2})^{-1} + n^{-2}\sum_{i=1}^{n}\sum_{j=1}^{n}\left\{\phi_{(2h^2+2g^2)^{1/2}} - 2\phi_{(h^2+2g^2)^{1/2}} + \phi_{(2g^2)^{1/2}}\right\}(X_i - X_j), \quad (4.24)$$

where $\phi_\sigma(\cdot)$ is the density of the normal distribution having mean zero and variance σ^2. Prior to evaluating (4.24), the pilot bandwidth g must be estimated. This is performed in a way similar to how this is done in the plug-in selector method (see Sect. 4.2.2). An example algorithm is presented in [188, Sect. 3.7].

4.2.5 *Other Selectors*

The four bandwidth selectors briefly presented above (ROT, LSCV, SCV and PI) are not the only ones described in the literature: one notes a number of less significant bandwidth selectors also being described. In addition, the four above mentioned groups have many variants, extensions and modifications. Nowadays, some of them are only of historical significance and are hardly ever being implemented. A description of many such methods can be found in [89, 181].

4.3 Multivariate Bandwidth Selectors

Multivariate KDE is an important technique in exploratory data analysis. The special attention is paid to the bivariate KDE, mainly because of the fact that it lends itself easily to visualization with familiar perspective or contour plots. As for high dimensional density functions, visualization techniques are not so obvious or natural and their interpretation encounters certain problems. The publication [158] shows some techniques of visualizing multivariate data but it seems that these techniques are not very popular among practitioners. We note that the multivariate KDE plays an important role in such statistical problems as nonparametric classification (clustering analysis, discriminant analysis), bump hunting, image segmentation and others similar, some of them briefly presented is Chap. 7.

Bivariate and multivariate KDE extensions are very similar from the analytical point of view and can be considered as a group. Many of the univariate bandwidth selection techniques presented in the former sections can be relatively easy extended to the multivariate case only if $\mathbf{H}$ is constrained to be a diagonal matrix ($\mathbf{H} = \text{diag}(h_1, \ldots, h_d)$) or a scalar multiplied by the identity matrix ($\mathbf{H} = h\mathbf{I_d}$). Many authors suggest that the later should not be used in practical applications with the only exception in case when the data pre-scaling is applicable (see an example is Sect. 3.4.1, Fig. 3.11). The unconstrained $\mathbf{H}$ has a fundamental advantage over the constrained counterparts as it permits arbitrary *shape* and *orientation* of the kernel function. This orientation could be done in an automatic fashion by using classical

KDE (3.26) generalized by the sample correlation matrix, but this approach is not valid in general, see Fig. 3.11 for an illustrative example.

4.3.1 Multivariate Rule-of-Thumb Selectors

As for the multivariate KDE, a rule-of-thumb selector was proposed in [23] and it has the following form (it was named by the authors as the *normal scale* selector)

$$\hat{\mathbf{H}}_{\mathrm{NS}} = \left(\frac{4}{n(d+2)}\right)^{2/(d+4)} \hat{\Sigma}, \tag{4.25}$$

where $\hat{\Sigma}$ is the estimate of the covariance matrix Σ and d is the problem dimensionality. Note also that in the case $d = 1$ the formula (4.25) is

$$\hat{h}_{\mathrm{NS}} = \left(\frac{4}{n(d+2)}\right)^{1/(d+4)} \hat{\sigma}, \tag{4.26}$$

and this formula coincides with (4.1). In the simplest case, if $\mathbf{H} = \mathrm{diag}(h_1, \ldots, h_d)$ and $\mathbf{\Sigma} = \mathrm{diag}(\sigma_1^2, \ldots, \sigma_d^2)$, the above can be written as

$$\hat{h}_{j,\mathrm{NS}} = \left(\frac{4}{n(d+2)}\right)^{1/(d+4)} \hat{\sigma}_j. \tag{4.27}$$

Yet another type of selector, which is less frequently used, was introduced in [177] and was dubbed the *maximal smoothing* selector. The name comes from the fact that this selector produces the smoothest KDEs consistent with the estimated scale of the data. This selector avoids the extreme sampling variability (a typical feature of cross validation methods, see Sect. 4.2.3). The disadvantage of this method is that it generates bandwidths that are not asymptotically optimal. This means that spurious features in the input data are not revealed. This selector has the following form

$$\hat{\mathbf{H}}_{\mathrm{MS}} = \left(\frac{(d+8)^{(d+6)/2}\, \pi^{d/2}\, R(K)}{16\, (d+2)\, n\, \Gamma(d/2+4)}\right)^{2/(d+4)} \hat{\Sigma}, \tag{4.28}$$

where Γ is the Gamma function [1] defined as

$$\Gamma(z) = \int_0^{\infty} x^{z-1} e^{-x}\, dx. \tag{4.29}$$

4.3.2 Multivariate Least Squares Cross Validation Selector

The scaling of the LSCV results from the univariate to the multivariate case is almost immediate [54]. The LSCV objective function is defined in the following way

$$\mathrm{LSCV}(\mathbf{H}) = \int_{\mathbb{R}^d} \hat{f}(\boldsymbol{x}, \mathbf{H})^2 d\boldsymbol{x} - 2n^{-1} \sum_{i=1}^{n} \hat{f}_{-i}(\boldsymbol{X}_i, \mathbf{H}), \tag{4.30}$$

where

$$\hat{f}_{-i}(\boldsymbol{x}, \mathbf{H}) = (n-1)^{-1} \sum_{j=1, j \neq i}^{n} K_{\mathbf{H}}(\boldsymbol{x} - \boldsymbol{X}_j), \tag{4.31}$$

is the *leave-one-out* estimator of f. In such a case, the LSCV objective function can be expressed as follows

$$\begin{aligned} \mathrm{LSCV}(\mathbf{H}) &= n^{-2} \sum_{i=1}^{n} \sum_{j=1}^{n} (K_{\mathbf{H}} * K_{\mathbf{H}})(\boldsymbol{X}_i - \boldsymbol{X}_j) \\ &\quad - 2n^{-1}(n-1)^{-1} \sum_{i=1}^{n} \sum_{j=1, j \neq i}^{n} K_{\mathbf{H}}(\boldsymbol{X}_i - \boldsymbol{X}_j), \end{aligned} \tag{4.32}$$

where $*$ denotes the convolution operator. Most practical implementations use the normal kernels, i.e. $K = \phi$, yielding $K_{\mathbf{H}} * K_{\mathbf{H}} = K_{2\mathbf{H}}$. The LSCV bandwidth matrix $\hat{\mathbf{H}}_{\mathrm{LSCV}}$ is the minimizer of the $\mathrm{LSCV}(\mathbf{H})$ objective function

$$\hat{\mathbf{H}}_{\mathrm{LSCV}} = \operatorname*{argmin}_{\mathbf{H} \in \mathcal{F}} \mathrm{LSCV}(\mathbf{H}), \tag{4.33}$$

and it has to be minimized numerically. In the first phase of the numerical optimization, an initial bandwidth has to be selected. Usually, it is enough to use the normal reference bandwidth (4.25) or the maximum smoothing bandwidth (4.28) to avoid certain initial spurious bumps, which can influence the optimization algorithm. Finally, after getting rid of the troublesome condition that $j \neq i$ in the second double summation (to make the FFT-based fast algorithm implementable, see Sect. 5.4.2), the objective function can be written as

$$\mathrm{LSCV}(\mathbf{H}) = n^{-2} \sum_{i=1}^{n} \sum_{j=1}^{n} T_{\mathbf{H}}(\boldsymbol{X}_i - \boldsymbol{X}_j) + 2n^{-1} K_{\mathbf{H}}(\mathbf{0}), \tag{4.34}$$

where

$$\begin{aligned} T_{\mathbf{H}}(u) &= (K_{\mathbf{H}} * K_{\mathbf{H}})(u) - 2K_{\mathbf{H}}(u) \\ &= K_{2\mathbf{H}}(u) - 2K_{\mathbf{H}}(u), \\ K_{\mathbf{H}}(\mathbf{0}) &= (2\pi)^{-d/2}|\mathbf{H}|^{-1/2}. \end{aligned} \tag{4.35}$$

There are two observations to be made. First, let us note that since $\mathbf{H}$ is a symmetric matrix, it is enough to optimize over only $\frac{1}{2}d(d+1)$ independent parameters, instead of d^2. Second, we emphasize that the goal here is to find the optimal $\mathbf{H}$, which has to be a *positive definite matrix*. Usually, there is no direct way of limiting the scope of the search performed by optimization software, like for example in the `stats::optim` and `stats::nlm` *R* functions [135], to the area of positive definite matrices (note that showing an *R* function, the usual notation `packageName::functionName` is used). Another problem is that many computer programs, including the `stats::optim` and `stats::nlm` functions, can only optimize over vectors, whereas the bandwidth optimization is done over the space of all positive definite matrices. So one possible trick is to let `stats::optim` search for the optimal x in d-dimensional Euclidean space but evaluate the objective function as a function of xx^T, i.e. $x_{\text{LSCV}} = \operatorname{argmin}_x \text{LSCV}(xx^T)$, so that the positive definite matrix that minimizes LSCV is thus $\{x_{\text{LSCV}}\ (x_{\text{LSCV}})^T\}$ [45].

4.3.3 Notation for Higher-Order Derivatives

Before presenting the multivariate version of the PI selector in Sect. 4.3.4, a special notation has to be introduced. This notation was first presented in [19] and was used subsequently by the authors in their follow-up articles (see the Reference section for more details).

Let f be a real d-variate function with its first ($r = 1$) derivative (*gradient*) vector defined as

$$\mathsf{D}f = \partial f/\partial \boldsymbol{x} = (\partial f/\partial x_1, \ldots, \partial f/\partial x_d), \tag{4.36}$$

with $\boldsymbol{x} = (x_1, \ldots, x_d)^T$. Therefore, all the second order partial derivatives ($r = 2$) can usually be organized into the *Hessian matrix*

$$\mathsf{H}f = \partial^2 f/(\partial \boldsymbol{x} \partial \boldsymbol{x}^T) = \partial^2 f/(\partial x_i \partial x_j)_{i,j=1,\ldots,d}, \tag{4.37}$$

and this matrix is of size $d \times d$. The *Hessian operator* can be formally written as $\mathsf{H} = \mathsf{D}\mathsf{D}^T$.

For $r \geq 3$ it is not clear how to organize the set containing all the d^r partial derivatives of order d into a matrix-like manner. One possible approach is to use the *equivalent vectorized* form where the r-th derivative of f is defined to be the *vector*

$$\mathsf{D}^{\otimes r} f = (\mathsf{D} f)^{\otimes r} = \partial^r f / \partial \boldsymbol{x}^{\otimes r} \in \mathbb{R}^{d^r}, \tag{4.38}$$

and this vector contains all partial derivatives of order r. In this notation, $\mathsf{D}^{\otimes r}$ denotes the r-th *Kronecker power* of the operator D, formally defined as the r-fold product $\mathsf{D} \otimes \cdots \otimes \mathsf{D}$. Naturally, $\mathsf{D}^{\otimes 0} f = f$, $\mathsf{D}^{\otimes 1} f = \mathsf{D} f$. For example, the equivalent vectorized form for the Hessian operator (that is if $r = 2$) is

$$\mathsf{D}^{\otimes 2} = \operatorname{vec} \mathsf{H}, \tag{4.39}$$

and of course

$$\mathsf{D}^{\otimes 2} f = \operatorname{vec} \mathsf{H}\, f \in \mathbb{R}^{d^2}. \tag{4.40}$$

Here 'vec' denotes the *vectorization operator*, which concatenates the columns of a matrix into a single vector [91], for example

$$\operatorname{vec} \begin{bmatrix} a & b \\ c & d \end{bmatrix} = \begin{bmatrix} a \\ c \\ b \\ d \end{bmatrix}. \tag{4.41}$$

For example for $r = 2$ and $d = 2$

$$\mathsf{D}^{\otimes 2} f = \left[\frac{\partial^2 f}{\partial x_1^2}, \frac{\partial^2 f}{\partial x_1 \partial x_2}, \frac{\partial^2 f}{\partial x_2^2}, \frac{\partial^2 f}{\partial x_2 \partial x_1} \right]^T. \tag{4.42}$$

It is obvious that this vectorization introduces some redundancy as for example (4.42) contains repeated mixed partial derivatives. Another possible vectorization $\operatorname{vech} \mathsf{H} f$ contains only the unique second order partial derivatives, where 'vech' operator is defined in (3.58). However, in the context of KDE-related problems the vec-based vectorization is preferred.

In cases when $r = 1$ and $r = 2$ the above vectorized form is not very useful, since the matrix form is commonly used but for $r > 2$ the $\mathsf{D}^{\otimes r}$ notation seems to be very convenient, clear and useful.

4.3.4 Multivariate Plug-In Selector

It was proved in [19] that AMISE defined as in (3.56) is not a good choice for deriving the multivariate PI unconstrained selector. A better formula would be

$$\begin{aligned}\text{AMISE}(\mathbf{H}) &\equiv \text{AMISE}\hat{f}(\cdot;\mathbf{H})\\ &= n^{-1}|\mathbf{H}|^{-1/2}R(K) + \frac{1}{4}\mu_2(K)^2(\text{vec}^T\mathbf{H})^{\otimes 2}\boldsymbol{\psi}_4. \end{aligned} \tag{4.43}$$

Note that the symbol $\boldsymbol{\Psi}$ in (3.56) does not stand for the same as a similar symbol $\boldsymbol{\psi}$ used in (4.43): the former represents a matrix, while the latter represents a vector. The use of the two very similar symbols, which can potentially lead to confusion, is dictated by the desire to remain consistent with the relevant references [19, 21, 22, 50].

The vector $\boldsymbol{\psi}_4$ is the only unknown quantity in (4.43). Comparing (3.56) and (4.43), it is easy to notice that the 'vech' operator was replaced by the 'vec' operator and this seemingly small change makes the matrix analysis more straightforward and in fact is a key strategic trick that makes it possible to construct the unconstrained multivariate PI selector.

It can be shown that the estimator of $\boldsymbol{\psi}_r$ is

$$\hat{\boldsymbol{\psi}}_r(\mathbf{G}_r) = n^{-2}\sum_{i=1}^{n}\sum_{j=1}^{n}\mathsf{D}^{\otimes r}L_{\mathbf{G}_r}(\boldsymbol{X}_i - \boldsymbol{X}_j), \tag{4.44}$$

for a given *even* number r based on a pilot bandwidth matrix $\mathbf{G}_r$. $\mathsf{D}^{\otimes r}L_{\mathbf{G}_r}$ is a vector containing all the r-th order partial derivatives of $L_{\mathbf{G}_r}$. This way $\hat{\boldsymbol{\psi}}_r(\mathbf{G}_r)$ is the vector of length d^r. Generally, L can be different that K but in practical implementations usually $K = L = \phi$.

In (4.43), $\boldsymbol{\psi}_4$ is unknown and, following the same multi-stage plug-in mechanism as in the univariate case, has to be estimated. In practical implementations (see for example [49]), the number of stages is typically set to be $l = 2$, thus the PI algorithm has the following form:

Step 1: Compute $\hat{\mathbf{G}}_8$ from the normal scale rule, set $r = 8$

$$\mathbf{G}_r^{NS} = \left(\frac{2}{n(r+d)}\right)^{2/(r+d+4)} 2\Sigma. \tag{4.45}$$

Step 2: Use $\hat{\mathbf{G}}_8$ to compute $\hat{\boldsymbol{\psi}}_8(\hat{\mathbf{G}}_8)$.

Step 3: Plug $\hat{\boldsymbol{\psi}}_8(\hat{\mathbf{G}}_8)$ into (4.46) and minimize to obtain $\hat{\mathbf{G}}_6$, set $r = 6$. In the first phase of the numerical optimization (using for example the `stats::optim` or `stats::nlm` *R* functions) an initial bandwidth has to be selected. Usually, it is enough to use the normal reference bandwidth (4.45), that is $\mathbf{G}_6^{NS}$.

$$\mathrm{AB}^2(\hat{\mathbf{G}}_r) = \left\| n^{-1}|\hat{\mathbf{G}}_r|^{-1/2}(\hat{\mathbf{G}}_r^{-1/2})^{\otimes r}\mathsf{D}^{\otimes r}L(0) + \frac{1}{2}\mu_2(L)(\mathrm{vec}^T \otimes \mathbf{I}_{d^r})\hat{\boldsymbol{\psi}}_{r+2}(\hat{\mathbf{G}}_{r+2}) \right\|^2 \tag{4.46}$$

($\mathsf{D}^{\otimes r}L(0)$ is a vector of all r-th partial derivatives of the normal density at $x = 0$. Its size is d^r).

Step 4: Use $\hat{\mathbf{G}}_6$ to compute $\hat{\boldsymbol{\psi}}_6(\hat{\mathbf{G}}_6)$.

Step 5: Plug $\hat{\boldsymbol{\psi}}_6(\hat{\mathbf{G}}_6)$ in (4.46) and minimize to obtain $\hat{\mathbf{G}}_4$, set $r = 4$. In the first phase of the numerical optimization (using for example the `stats::optim` or `stats::nlm` *R* functions) an initial bandwidth must be selected. Usually, it is enough to use the normal reference bandwidth (4.45), that is $\mathbf{G}_4^{NS}$.

Step 6: Use $\hat{\mathbf{G}}_4$ to compute $\hat{\boldsymbol{\psi}}_4(\hat{\mathbf{G}}_4)$. s

Step 7: Finally, plug $\hat{\boldsymbol{\psi}}_4$ into (4.43) and numerically minimize to obtain the bandwidth value.

$$\hat{\mathbf{H}}_{\mathrm{PI},2} = \underset{\mathbf{H}\in\mathcal{F}}{\operatorname{argmin}}\ \mathrm{AMISE}(\mathbf{H}). \tag{4.47}$$

The above algorithm steps are taken directly from [19]. However, a number of experiments carried out by us clearly show that in vast majority of the datasets, the first three steps are not necessary and $\hat{\mathbf{G}}_6$ can be replaced with the normal scale approximation using (4.45), that is $\mathbf{G}_6^{NS}$ with the rest of the steps being the same. The accuracy of both the complete algorithm and the reduced one (measured using the ISE criterion (3.53)) is practically the same.

Note also that in [19] the above algorithm is generalized taking into account any number of stages l. However, it seems that more research is needed to formally prove the correctness of the algorithm and to find realistic examples where $l > 2$ is a natural choice.

4.3.5 Multivariate Smoothed Cross Validation Selector

The unconstrained SCV bandwidth selector was introduced in [20]. The multivariate counterpart of the univariate objective function (4.24) is

$$\mathrm{SCV}(\mathbf{H}) = n^{-1}|\mathbf{H}|^{-1/2}(4\pi)^{-1/2} + n^{-2}\sum_{i=1}^{n}\sum_{j=1}^{n}\left\{\phi_{(2\mathbf{H}+2\mathbf{G})} - 2\phi_{(\mathbf{H}+2\mathbf{G})} + \phi_{2\mathbf{G}}\right\}(\boldsymbol{X}_i - \boldsymbol{X}_j). \tag{4.48}$$

G is a *pilot bandwidth* matrix that has to be estimated in a similar way as in the case of the PI selectors. The only difference is that Step 6 is not necessary, as in (4.48) the $\hat{\mathbf{G}}_4$, and not $\hat{\boldsymbol{\psi}}_4$, is required. Finally, in Step 7 one needs to plug $\hat{\mathbf{G}}_4$ into (4.48) and numerically minimize to obtain the bandwidth value.

$$\hat{\mathbf{H}}_{\mathrm{SCV},2} = \underset{\mathbf{H}\in\mathcal{F}}{\operatorname{argmin}}\, \mathrm{SCV}(\mathbf{H}). \tag{4.49}$$

4.3.6 Simulations for a Bivariate Target Densities

Let us finish the description of the three main multivariate bandwidth selectors (PI, SCV, LSCV/UCV) with a visual comparison of their performance. The bivariate target densities to be analyzed were taken from [18] as these cover a very wide range of density shapes. Their original names and numbering are preserved. The shapes are shown in Fig. 4.3.

For each target density, 100 replicates were drawn and the ISE errors (3.53) were computed for every bandwidth selector. Then, classical boxplots were drawn and the results are shown in Fig. 4.4. It can be easily noticed that the PI selector usually outperforms the two remaining selectors, but the differences between the PI and the SCV are very slight. On the other hand, the high variability of the LSCV/UCV selector is also quite clear, making this selector not really fit for practical applications.

4.4 Computational Aspects

As for finding the optimal bandwidth, the computational problems involved are more difficult compared with the problems related to pure KDE computations. Typically, to complete all the required computations in terms of this task, a sort of numerical optimization is needed over an objective function (see the PI, SCV and UCV methods described in earlier sections of this chapter). Usually, the computational complexity of evaluating typical objective function is $O(n^2)$. In most cases, the objective functions have to be evaluated many times (often more than a hundred or so), making the problem of finding the optimal bandwidth very computationally expensive, even for moderate data dimensionalities and sizes.

PI, SCV and UCV methods rely on evaluating equations of the type (4.3) for the univariate case and of the type (4.44) for the multivariate case (i.e., the integrated density derivative functionals). Fast and accurate computation of these functionals

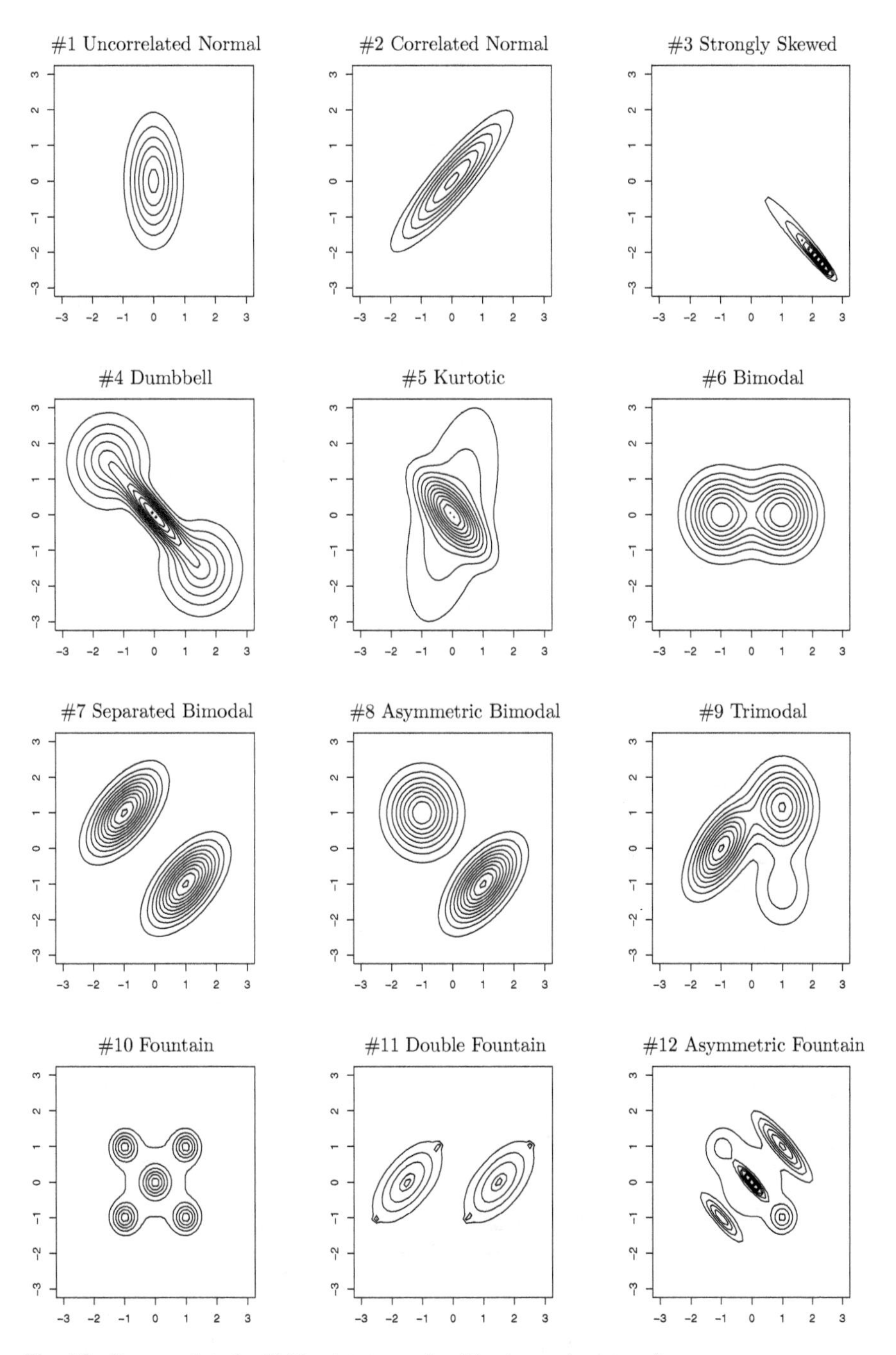

Fig. 4.3 Contour plots for 12 bivariate target densities (normal mixtures)

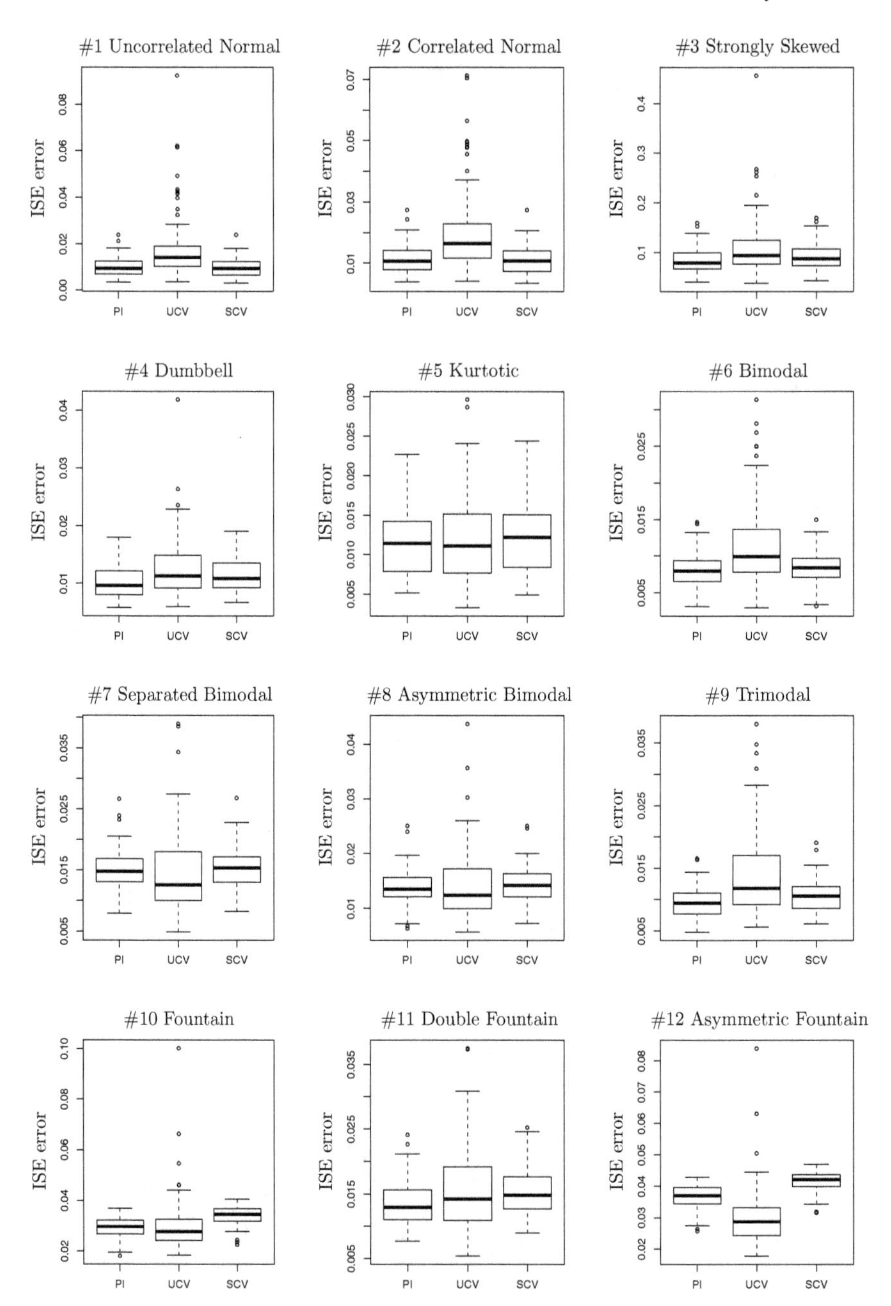

Fig. 4.4 Boxplots of the ISE error for the KDE using the three bandwidth selectors (PI, SCV, LSCV/UCV) for the bivariate target densities and for the sample size $n = 100$

is a crucial issue for bandwidth selection. Note also that it is only the PI and SCV selectors that involve derivatives (at least of order 4) and this extension makes the problem of fast bandwidth selection really complex and time consuming. LSCV/UCV also involves computation of (4.3) and (4.44) but in its simplest form where $r = 0$.

Chapter 5 is devoted in its entirety to this extremely important problem. There, a FFT-based algorithm is presented in details. This algorithm assumes a type of data pre-discretization known as binning (see Sect. 5.2). There are other approaches aimed at fast bandwidth computation based on e.g. utilizing such hardware-based techniques as Field-Programmable Gate Arrays (FPGA) [73], see Chap. 6. Note however that in the case of FPGA, only preliminary results are presented. We have fully implemented only the univariate plug-in bandwidth selector presented is Sect. 4.2.2. We emphasize that the results obtained are very promising and the speedups are really significant.

Chapter 5
FFT-Based Algorithms for Kernel Density Estimation and Bandwidth Selection

5.1 Introduction

Analyzing individual formulas for KDE (Chap. 3) and bandwidth selection (Chap. 4), it is not difficult to notice that $O(n^2)$ computational complexity is a common case. Consequently, for large datasets, naive computations are a very bad decision. As for now, there research on fast KDE and bandwidth computations is not a large endeavor. This chapter is a detailed description of a very elegant and effective method based on Fast Fourier Transform (FFT), which allows a huge computational speedups without a loss of accuracy. The results presented in this Chapter can be considered as a broad extension of the results given in [186], especially if used for multivariate KDE and multivariate unconstrained bandwidth selectors. The material is based mainly on our results presented in [71, 72].

The chapter is organized as follows: Sect. 5.2 presents the concept of data binning, which can be understood as a kind of data discretization. Binning data is a must-have step before applying the FFT-based technique for fast KDE computation. Section 5.3 is devoted to a detailed description of this technique. Section 5.4 shows how the FFT-based algorithm can be extended to support algorithms for bandwidth selection for KDE. Section 5.5 contains computer simulation results showing the speed of the FFT-based algorithm used for KDE computations. Section 5.6 contains computer simulation results, with the accuracy and speed of the algorithm for bandwidth selection being the focus of analysis.

5.2 Data Binning

Most commonly, *binning* is described as a way of grouping a number of more or less continuous values into a smaller number of *bins*. For example, let us consider data describing a group of people. One might want to sort the people by age into bins

A. Gramacki, *Nonparametric Kernel Density Estimation and Its Computational Aspects*, Studies in Big Data 37,
https://doi.org/10.1007/978-3-319-71688-6_5

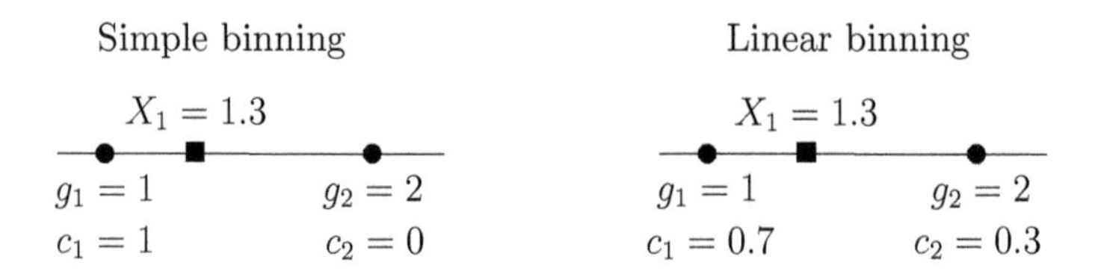

Fig. 5.1 The idea of simple and linear binning

representing a smaller number of age ranges. The well-known histogram is a typical example of data binning. In this book the notion of binning is understood in a slightly different way. By binning we mean a kind of data *discretization* (or approximation) when the original dataset (uni- or multivariate) is replaced by a regular grid of points and grid weights. The two most common *binning rules* are: *simple binning* and *linear binning*. The latter is much more often used in practice. The reader is invited to consult [85, 155, 161, 167] for univariate case. The multivariate case with the required mathematical background was presented in [186, 188], therefore we do not describe it here. The problem of binning accuracy is analyzed in [69, 80].

5.2.1 Univariate Case

The task here is to find, on the basis of n data points $X_1, X_2, \ldots, X_n$ (experimental data), a new set of M *equally spaced grid points* $g_1, g_2, \ldots, g_M$ with associated *grid counts* $c_1, c_2, \ldots, c_M$ (in practical applications, a usual requirement is to have $M \ll n$). The grid counts are obtained by assigning certain weights to the grid points, based on neighboring observations. In other words, each grid point is accompanied with a corresponding grid count. Consider a simple binning procedure: if a data point at X_1 has surrounding grid points g_1 and g_2, then the process involves assigning a grid count of 1 to the grid point closest to X_1. This idea is depicted in Fig. 5.1 (left picture). In this simple example only one data point is here (filled square) as well as two grid points (two filled circles). In this example, $c_1 = 1$ as X_1 is closer to g_1 as to g_2.

In linear binning (Fig. 5.1, right picture), grid counts are assigned for both grid points according to the following rules

$$
\begin{aligned}
c_1 &= \frac{g_2 - X_1}{g_2 - g_1}, \\
c_2 &= \frac{X_1 - g_1}{g_2 - g_1},
\end{aligned}
\tag{5.1}
$$

where $c_1 + c_2 = n = 1$.

Below, we present a more extensive example of the linear binning (with $n = 9$ and $M = 4$)

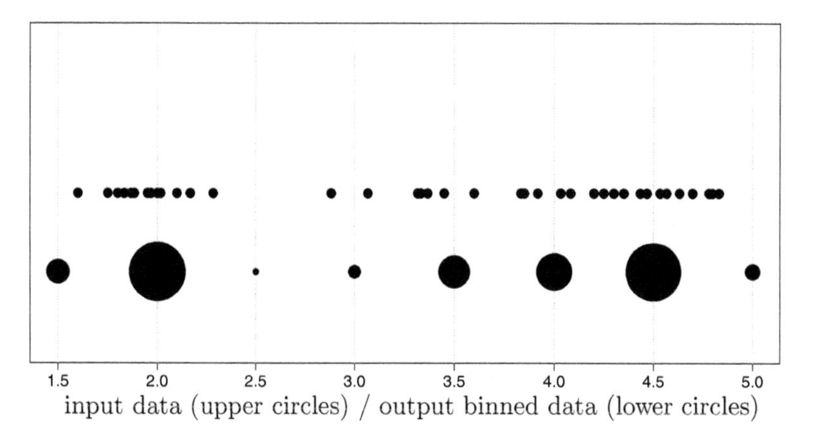

Fig. 5.2 Visualization of the univariate linear binning. $n = 50$ input data points are replaced by $M = 8$ grid points with appropriate grid counts assigned labelfig

$$
\begin{aligned}
X_i &= \{0,\ 0.1,\ 0.3,\ 0.3,\ 0.3,\ 1,\ 2.3,\ 2.7,\ 3\}, \quad i = 1, \ldots, n = 9, \\
g_j &= \{0,\ 1,\ 2, 3\}, \quad j = 1, \ldots, M = 4, \\
c_j &= \{4,\ 2,\ 1,\ 2\}, \quad \sum_{j=1}^{M} c_j = n.
\end{aligned} \tag{5.2}
$$

Figure 5.2 shows the linear binning for a sample dataset $X_i,\ i = \{1, \ldots, 50\}$ (see the *R* system printout below). The upper line of filled circles represent the input dataset, while the lower line of filled circles represent grid counts $c_j,\ j = \{1, \ldots, 8\}$ located in the equally spaced grid points $g_j,\ j = \{1, \ldots, 8\}$. The diameters of the lower line circles are proportional to c_j. It is easily to notice that the circles become bigger as the local concentration of input 'mass' grows.

```
> X
 [1] 3.600 1.800 3.333 2.283 4.533 2.883 4.700 3.600 1.950 4.350 1.833
[12] 3.917 4.200 1.750 4.700 2.167 1.750 4.800 1.600 4.250 1.800 1.750
[23] 3.450 3.067 4.533 3.600 1.967 4.083 3.850 4.433 4.300 4.467 3.367
[34] 4.033 3.833 2.017 1.867 4.833 1.833 4.783 4.350 1.883 4.567 1.750
[45] 4.533 3.317 3.833 2.100 4.633 2.000
> min(X); max(X)
[1] 1.6
[1] 4.833
> g
[1] 1.5 2.0 2.5 3.0 3.5 4.0 4.5 5.0
> c
[1]  4.934 11.932  1.368  2.698  6.602  7.534 11.702  3.230
> sum(c)
[1] 50
```

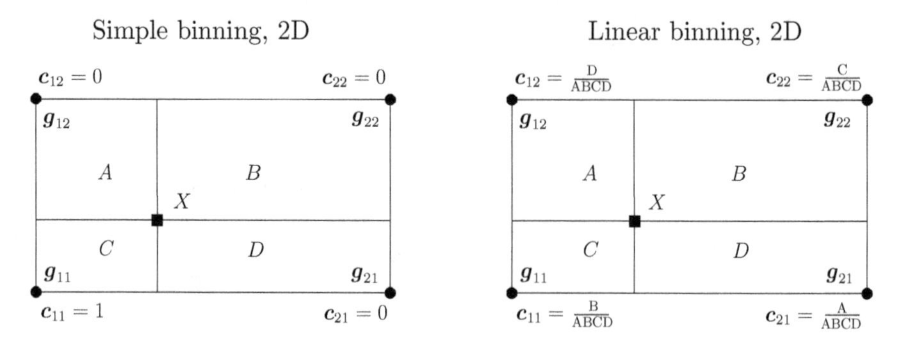

Fig. 5.3 The ideas behind simple and linear binning, bivariate case

5.2.2 Multivariate Case

The idea behind both simple and linear binning can obviously be extended to cover the multivariate case. The following notation is used [186]: let g be equally spaced grid points and $\boldsymbol{c}$ be grid counts. For $k = 1, \ldots, d$, let $g_{k1} < \cdots < g_{kM_K}$ be an equally spaced grid in the kth coordinate directions such that $[g_{k1}, g_{kM_k}]$ contains the kth coordinate grid points. Here M_k is a positive integer representing the *grid size* in direction k. Let

$$g_i = (g_{1i_1}, \ldots, g_{di_d}), \quad 1 \le i_k \le M_k, \quad k = 1, \ldots, d, \tag{5.3}$$

denote the grid point indexed by $\boldsymbol{i} = (i_1, \ldots, i_d)$ and the kth *binwidth* (or *mesh size*) be denoted by

$$\delta_k = \frac{g_{kM_k} - g_{k1}}{M_k - 1}. \tag{5.4}$$

Te same notation is used for indexing $\boldsymbol{c}$.

Figure 5.3 is a graphical presentation (for bivariate case, that is when $d = 2$) of how a sample data point X distributes its weight to neighboring grid points. The picture is taken from [186]. For simple binning, the point X gives all of its weight to its nearest grid point. In this example, the lower left grid point takes the weight equal to 1, that is $\boldsymbol{c}_{11} = 1$. In the case of linear binning, the contribution from X is distributed among each of the four surrounding grid points according to areas of the opposite sub-rectangles induced by the position of the point. The extension to higher dimensions is obvious with *volumes* being used instead of *areas*.

The number of grid points can of course be different in every dimension. Figure 5.4 shows an another toy example where the grid size is two in the first dimension and three in the second dimension. Two input data points (2, 2), (4, 3), marked by the filled squares, are replaced by six ($M_1 = 3$, $M_2 = 2$) grid points

$$(1, 1), (3, 1), (5, 1), (1, 5), (3, 5), (5, 5), \tag{5.5}$$

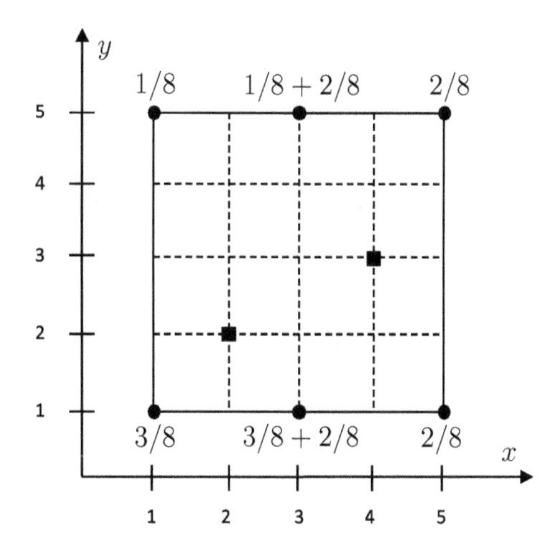

Fig. 5.4 Visualization of the bivariate linear binning

marked with the filled circles. As the result, six grid counts are obtained

$$3/8,\ (3/8+2/8),\ 2/8,\ 1/8,\ (1/8+2/8),\ 2/8, \tag{5.6}$$

and they obviously sum up to 2 (as $n = 2$).

Another example is based on a sample bivariate dataset *Unicef*, available from the *ks R* package, [182]), see Fig. 5.5. This data set describes the numbers of deaths of children under 5 years of age per 1000 live births and the average life expectancy (in years) at birth for 73 countries with the GNI (Gross National Income) less than 1000 US dollars per annum per capita. A data frame consists of 73 observations on 2 variables. Each observation corresponds to a country.

Each data point, shown as small filled circles, is replaced by a grid of *equally spaced grid points* (gray filled circles) of sizes 5×8, 10×10 and 20×20—see Figs. 5.5a, b and c, respectively. The actual values of grid counts $\boldsymbol{c}_j$ are proportional to the gray circle diameters. Grid counts are bigger for corresponding grid points that are located in the area of a bigger concentration of $\boldsymbol{x}$. Grid points with the corresponding grid counts equal to 0 are represented as small gray open squares. Note also that some gray filled circles are very small and almost invisible on the plots.

Figure 5.5d shows the relationship between the grid size and the percentage of non-zero grid points for the sample *Unicef* dataset. For the sake of simplicity, the assumption $M_1 = M_2$ was made but in general these values do not need to be the same. It can be easily observed that the percentage of non-zero grid points declines very rapidly, as the grid size increases. This phenomena can be also easily noticed when one compares Figs. 5.5a, b and c.

Figure 5.5e shows the contour plot for 5×8 grid mesh. Obviously, such a small mesh is very inaccurate in terms of practical applications. On the other hand, Fig. 5.5f shows the contour plot for 100×100 grid mesh that seems to be absolutely sufficient in terms of bivariate contour plots.

Below, the numerical results of binning from Fig. 5.5a are shown. Here $M_1 = 5$, $M_2 = 8$ (see (5.3)) and

- the grid points $g_{11}, \ldots, g_{15}$ in the first direction are {19.0, 93.2, 167.5, 241.8, 316.0},
- the grid points $g_{21}, \ldots, g_{28}$ in the second direction are {39.0, 43.9, 48.7, 53.6, 58.4, 63.3, 68.1, 73.0},
- the matrix of grid counts $\boldsymbol{c}_j$ ($\sum \boldsymbol{c}_j = 73 = n$) is

```
[,1]  [,2] [,3] [,4]  [,5]  [,6]  [,7]  [,8]
[1,] 0.00  0.00  0.0  0.0  0.35  1.9  6.470  4.87
[2,] 0.23  1.81  2.1  5.2  5.71  5.8  4.233  1.48
[3,] 2.80  5.54  6.0  6.2  0.95  0.2  0.092  0.13
[4,] 1.22  3.30  2.5  2.0  0.00  0.0  0.000  0.00
[5,] 0.83  0.74  0.2  0.0  0.00  0.0  0.000  0.00
```

5.3 The FFT-Based Algorithm for Density Estimation

5.3.1 *Introductory Notes and a Problem Demonstration*

This section is concerned with a FFT-based algorithm for kernel density estimation that was originally described in [186]. From now on this method is referred to as *Wand's algorithm*.

Although Wand's algorithm is very fast and accurate, it suffers from a serious limitation. It supports only a small subset of all possible multivariate kernels of interest. Two commonly used kernel types are *product* (3.24) and *radial*, also known as *spherically symmetric*, (3.25) ones. The problem appears when the radial kernel is used and the bandwidth matrix $\mathbf{H}$ is *unconstrained*, that is $\mathbf{H} \in \mathcal{F}$, where $\mathcal{F}$ denotes the class of symmetric, positive definite $d \times d$ matrices. If, however, the bandwidth matrix belongs to a more restricted *constrained* (*diagonal*) form (that is $\mathbf{H} \in \mathcal{D}$), the problem does not manifest itself.

To the best of our knowledge, the above-mentioned problem was not given a clear presentation and satisfying solution in the literature, except for a few short mentions in [186, 188] and in the `ks::kde` *R* function [49]. Moreover, many authors cite the FFT-based algorithm for KDE uncritically without discussing and taking into account some of its greatest limitations.

Note that Wand's algorithm is implemented in the *ks R* package [49], as well as in the *KernSmooth R* package [191]. However, the *KernSmooth* implementation only supports product kernels. The standard `stats::density` *R* function uses FFT to only compute the univariate kernel density estimates.

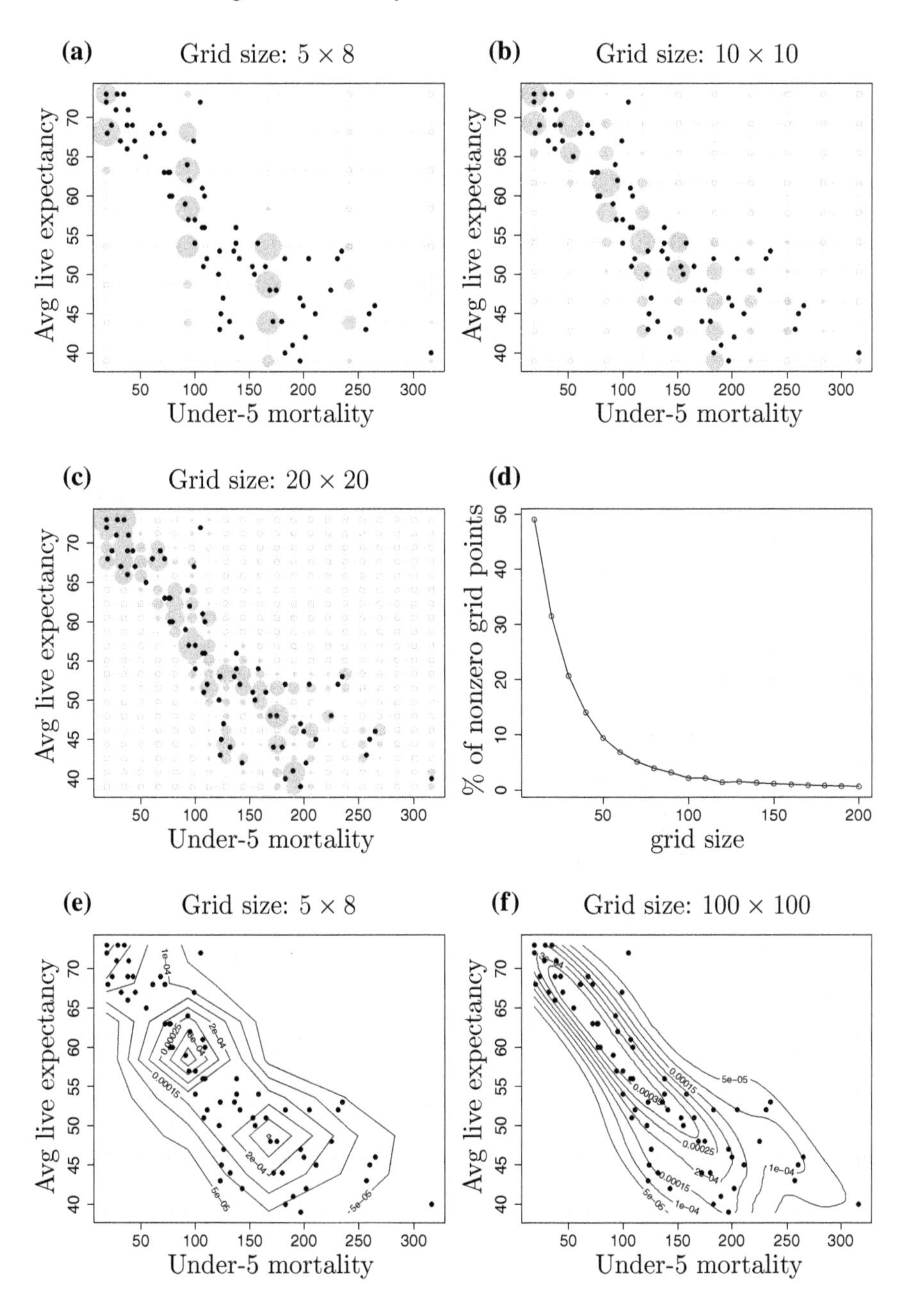

Fig. 5.5 An illustrative example of data binning. A sample 2D dataset *Unicef* is binned

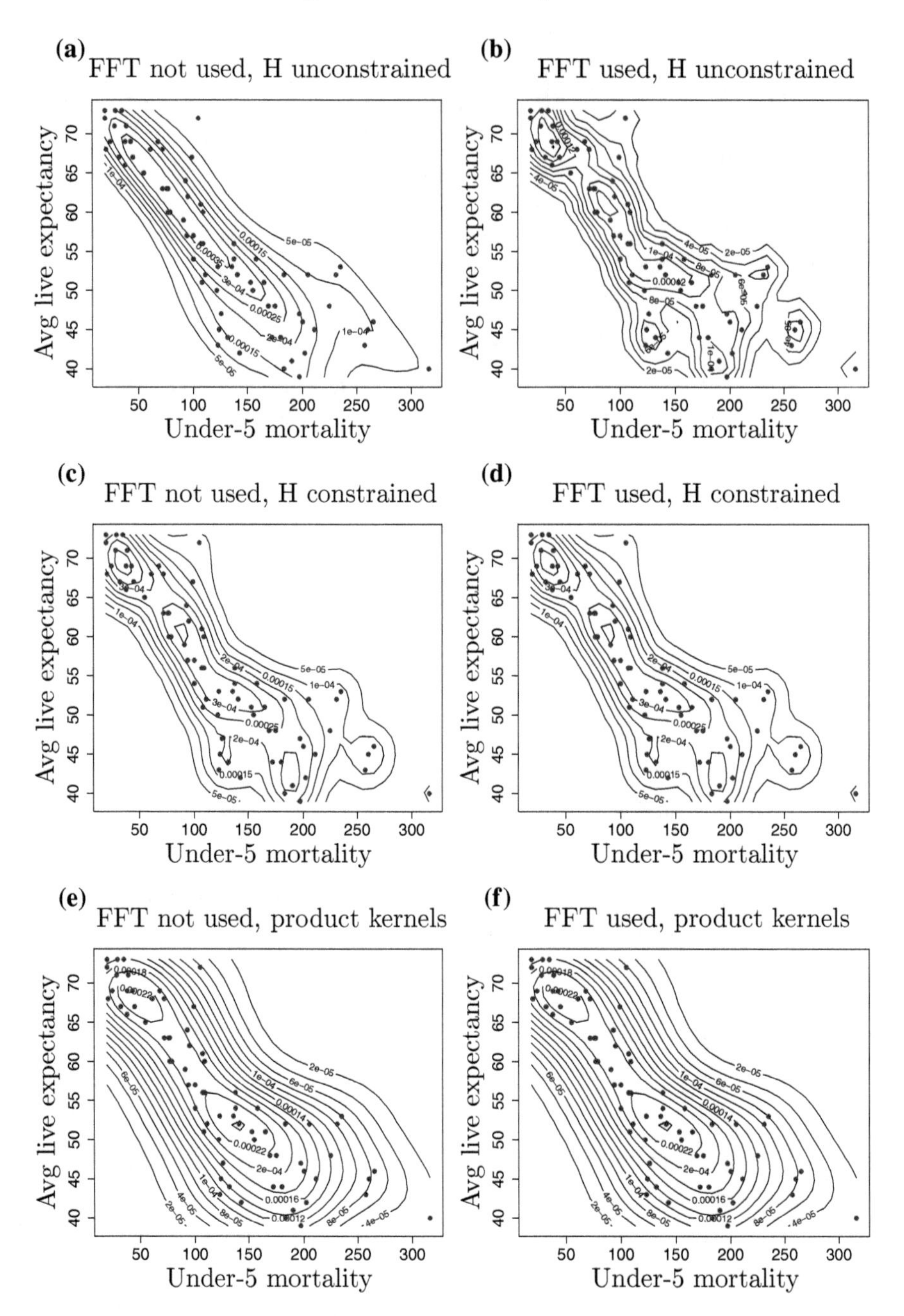

Fig. 5.6 Density estimates for a sample dataset with FFT and without it, for both unconstrained and constrained bandwidth matrices

For the sake of simplicity, we only present a 2D example but the same conclusions are valid also for higher dimensions. Figure 5.6 demonstrates the problem. The sample *Unicef* dataset from the *ks R* package was used. The density estimation depicted in Fig. 5.6a can be treated as a reference. It was calculated directly from (3.15). The unconstrained bandwidth **H** was calculated using the `ks::Hpi` *R* function. In Fig. 5.6b, the density estimation was calculated using Wand's algorithm. The bandwidth **H** was also unconstrained, as in to the case presented in Fig. 5.6a. It is easy to notice that the result is obviously inaccurate, as the results in Figs. 5.6a, b should be practically the same (only very small disparities may occur, typically due to the differences in FFT implementations, the inaccuracy of the floating-point arithmetic etc.). Figures 5.6c, d are the analogs of Figs. 5.6a, b, respectively, with the difference being that the bandwidth **H** is now constrained (calculated using the `ks::Hpi.diag` *R* function). Both figures are identical and, moreover, very similar to Fig. 5.6b. This similarity suggests that Wand's algorithm in some way loses most (or even all) of the information carried by the off-diagonal entries of the bandwidth matrix **H**. In other words, Wand's algorithm (in it's original form) is adequate only for constrained bandwidths.

Additionally, Figs. 5.6e, f show the results where the *product* kernel was used. Two individual scalar bandwidths were calculated using the `ks::hpi` *R* function. Note that the `ks::hpi` function (in lower-case) implements the univariate plug-in selector, while the `ks::Hpi` function (in upper-case) is its multivariate equivalent. Both figures are identical as the problem presented in here does not affect product kernels.

The above-mentioned problem has been successfully solved and the solution is presented in Sects. 5.3.2 and 5.3.3.

5.3.2 Univariate Case

As it was mentioned earlier, the computational complexity of (3.5) is $O(n^2)$, if the evaluation points x are the same as the input sample X_i. Here, for convenience this equation, which serves as a starting point, is reproduced

$$\hat{f}(x, h) = \frac{1}{n} \sum_{i=1}^{n} K_h(x - X_i). \tag{5.7}$$

In practical applications, it is more desirable to compute KDE for equally spaced grid points g_j where $j = 1, \ldots, M$, especially if n is very big, e.g. takes on the value in the range of thousands or even higher. For the univariate case, $M \approx 400$–500 seems to be absolutely sufficient in terms of most of the applications (note that quite often some lower values, say 100–200, also seems to be sufficient). Equation (5.7) can be obviously rewritten as

$$\hat{f}_j \equiv \hat{f}(g_j, h) = \frac{1}{n} \sum_{i=1}^{n} K_h\left(g_j - X_i\right), \quad j = 1, \ldots, M, \tag{5.8}$$

and now the number of kernel evaluations is $O(nM)$. For a large n, this value can grow too big for practical use. So, the next natural step could be to make use of binning, that is for every sample point X_i to be replaced by a pair of two values: the *grid point* g_i and the *grid count* c_i, as it was explained in Sect. 5.2.1. Thus we obtain

$$X_i \rightarrow \{g_i, c_i\}. \tag{5.9}$$

In that case, (5.7) can be again rewritten as below. Note also that now, $\tilde{f}$ instead of $\hat{f}$ was used as (5.10) is, in a way, an approximation of (5.7).

$$\tilde{f}_j \equiv \tilde{f}(g_j, h, M) = \frac{1}{n} \sum_{l=1}^{M} K_h\left(g_j - g_l\right) c_l, \quad j = 1, \ldots, M. \tag{5.10}$$

Now, the number of kernel evaluations is $O(M^2)$. If, however, the grid points are equally spaced (which is in practice almost always true, not equally spaced gridding is simply impractical), then the number of kernel evaluations is $O(0.5\ M(M+1))$. This is because the kernel is a symmetric function and thus $K_h\left(g_j - g_l\right) = K_h\left(g_l - g_j\right)$. Therefore, since these two values are the same, it is enough to calculate this value only once. But the number of multiplications $K_h(\cdot)\, c_l$ is still $O(M^2)$. To reduce this value to $O(M \log_2 M)$, the FFT-based technique can be used. This technique is presented in the subsequent parts of this chapter.

To use the FFT for a fast computation of (5.10), this equation must be rewritten again as

$$\begin{aligned} \tilde{f}_j &= \frac{1}{n} \sum_{l=1}^{M} K_h\left(g_j - g_l\right) c_l, \\ &= \sum_{l=1}^{M} k_{j-l} c_l, \end{aligned} \tag{5.11}$$

where $j = 1, \ldots, M$ and

$$\begin{aligned} k_{j-l} &= \frac{1}{n} K_h(\delta(j-l)), \\ \delta &= \left(\frac{b-a}{M-1}\right). \end{aligned} \tag{5.12}$$

Here, $a = g_1, b = g_M$ and δ is the grid width. For the toy example given by (5.2), it is clear that $\delta = (3-0)/(4-1) = 1$. For the example from Fig. 5.2, $\delta = (4.833 - 1.6)/(8-1) = 0.462$. The second summation in (5.11) has not yet had the form of the 'pure' convolution. The goal is to get a convolution-like equation, which can be solved very fast using the FFT algorithm. In order to represent it as convolution, we should first observe that

$$\begin{aligned} & c_l = 0 \quad \text{for} \quad l \quad \text{not in the set} \quad \{1, \dots, M\}, \\ & \text{and} \\ & K(-x) = K(x). \end{aligned} \tag{5.13}$$

In that case, the summation can be safely extended to $-M$, that is

$$\tilde{f}_j = \sum_{l=-M}^{M} k_{j-l} c_l = \boldsymbol{c} * \boldsymbol{k}, \tag{5.14}$$

where $*$ is the convolution operator. Now, (5.14) has the form of the classical discrete convolution of $\boldsymbol{k}$ and $\boldsymbol{c}$. For a toy example when $M = 3$ and $j = \{1, 2, 3\}$ the summation in (5.14) takes the following form:

$$\begin{aligned} & \tilde{f}_{j=1} : \underline{k_4 c_{-3}} + \underline{k_3 c_{-2}} + \underline{k_2 c_{-1}} + \underline{k_1 c_0} + k_0 c_1 + k_{-1} c_2 + k_{-2} c_3, \\ & \tilde{f}_{j=2} : \underline{k_5 c_{-3}} + \underline{k_4 c_{-2}} + \underline{k_3 c_{-1}} + \underline{k_2 c_0} + k_1 c_1 + k_0 c_2 + k_{-1} c_3, \\ & \tilde{f}_{j=3} : \underline{k_6 c_{-3}} + \underline{k_5 c_{-2}} + \underline{k_4 c_{-1}} + \underline{k_3 c_0} + k_2 c_1 + k_1 c_2 + k_0 c_3. \end{aligned} \tag{5.15}$$

Given (5.13), we see that the underscored expressions are zeroed out. As the convolution operator is commutative, (5.14) can be rewritten as

$$\tilde{f}_j = \sum_{l=-M}^{M} c_{j-l} k_l, \tag{5.16}$$

where

$$k_l = \frac{1}{n} K_h(\delta l). \tag{5.17}$$

Now, the summation in (5.16) is defined as follows:

$$\begin{aligned} & \tilde{f}_{j=1} : \underline{c_4 k_{-3}} + c_3 k_{-2} + c_2 k_{-1} + c_1 k_0 + \underline{c_0 k_1} + \underline{c_{-1} k_2} + \underline{c_{-2} k_3}, \\ & \tilde{f}_{j=2} : \underline{c_5 k_{-3}} + \underline{c_4 k_{-2}} + c_3 k_{-1} + c_2 k_0 + c_1 k_1 + \underline{c_0 k_2} + \underline{c_{-1} k_3}, \\ & \tilde{f}_{j=3} : \underline{c_6 k_{-3}} + \underline{c_5 k_{-2}} + \underline{c_4 k_{-1}} + c_3 k_0 + c_2 k_1 + c_1 k_2 + \underline{c_0 k_3}, \end{aligned} \tag{5.18}$$

obviously, obtaining the same result as by using (5.15).

The observation that in both (5.15) and (5.18) the factors for $l = -M$ and $l = M$ are always zeroed out helps to slightly reduce the computational burden. Consequently, the final form for $\tilde{f}_j$ is

$$\tilde{f}_j = \sum_{l=-(M-1)}^{M-1} c_{j-l} k_l. \tag{5.19}$$

Now, the summation in (5.19) is as follows:

$$\begin{aligned}
\tilde{f}_{j=1} &: c_3 k_{-2} + c_2 k_{-1} + c_1 k_0 + \underline{c_0 k_1} + \underline{c_{-1} k_2}, \\
\tilde{f}_{j=2} &: \underline{c_4 k_{-2}} + c_3 k_{-1} + c_2 k_0 + c_1 k_1 + \underline{c_0 k_2}, \\
\tilde{f}_{j=3} &: \underline{c_5 k_{-2}} + \underline{c_4 k_{-1}} + c_3 k_0 + c_2 k_1 + c_1 k_2.
\end{aligned} \tag{5.20}$$

Finally, the following vectors are involved in the computations

$$\begin{aligned}
\boldsymbol{c} &= (c_1, c_2, \ldots, c_M), \\
\boldsymbol{k} &= (k_{-(M-1)}, \ldots, k_{-1}, k_0, k_1, \ldots, k_{M-1}).
\end{aligned} \tag{5.21}$$

In (5.19) the vector $\tilde{f}_j$ has the character of a *discrete convolution* of two vectors $\boldsymbol{c}$ and $\boldsymbol{k}$ and can be calculated very effectively using the well-known FFT algorithm and the *discrete convolution theorem* [173] which states that under suitable conditions the Fourier transform of a convolution is the point-wise product of Fourier transforms. In other words, convolution in one domain (e.g. in the time domain) equals point-wise multiplication in the other domain (e.g. in the frequency domain), that is

$$f * g = \mathcal{F}^{-1}\{\mathcal{F}(f) \cdot \mathcal{F}(g)\}, \tag{5.22}$$

where $\mathcal{F}$ denotes the Fourier transform operator and $\mathcal{F}^{-1}$ is its inverse. To calculate $\tilde{f}_j$, the following three steps are required:

Step 1: Calculate the grid counts $cl, l = 1,\ldots, M$.

Step 2: Calculate the kernel weights kl for such values of l where $kl = 0$.

Step 3: Calculate the convolution (5.19).

The discrete convolution theorem places certain requirements on the form of vectors $\boldsymbol{c}$ and $\boldsymbol{k}$. In digital signal processing (DSP), the first is usually called the *input signal* and the second is called the *output signal*. First, it is assumed that $\boldsymbol{c}$ is a periodic signal; second, it is assumed that both $\boldsymbol{c}$ and $\boldsymbol{k}$ have the same length. If the two lengths are not the same (as it is here, see (5.21)), the special procedure known as *padding the signal with zeros* (often abbreviated as *zero-padding*) needs to be employed.

The first step is to define the length of the new vectors. As most computer implementations of the FFT algorithm work best when the input vectors are the powers of two, the new length should be defined such that $P = 2^k$ $(k \in \mathbb{N}_+)$ and $P \geq (3M - 1)$,

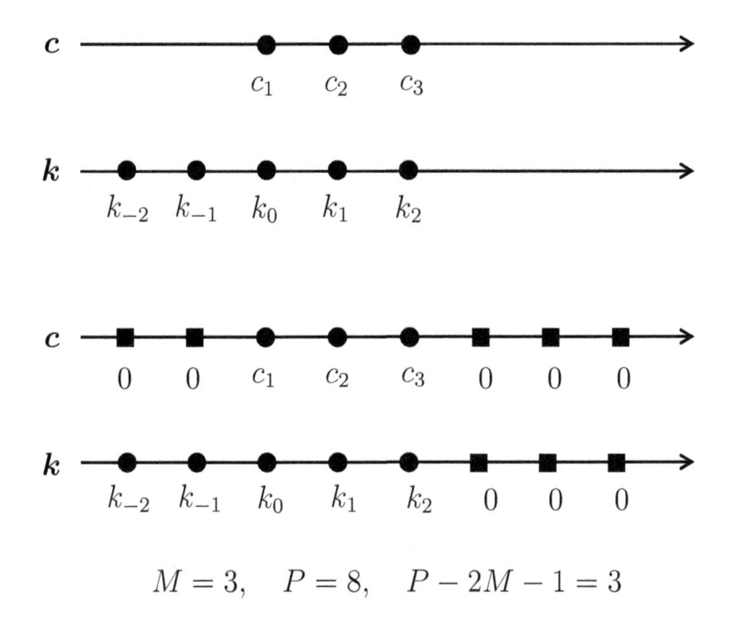

Fig. 5.7 Original vectors $\boldsymbol{c}$ and $\boldsymbol{k}$ (two upper axes) and the same vectors after zero-padding (two lower axes)

or equivalently

$$P = 2^{\lceil \log_2(3M-1) \rceil}, \tag{5.23}$$

where $\lceil x \rceil$ is the least integer greater than or equal to x.

If the P value is determined, the following scheme for zero-padding is proposed

$$\begin{aligned} \boldsymbol{c}_{zp} &= (\boldsymbol{0}_{M-1}, c_1, c_2, \ldots, c_M, \boldsymbol{0}_{P-2M-1}) \\ \boldsymbol{k}_{zp} &= (k_{-(M-1)}, \ldots, k_{-1}, k_0, k_1, \ldots, k_{M-1}, \boldsymbol{0}_{P-2M-1}). \end{aligned} \tag{5.24}$$

It is also worth noting that the proposed solution in the form of (5.24) is different than the one presented in [71, 188, Appendix D]. In the univariate case, the two approaches give exactly the same results. However, the multivariate extension based on [188] works well only if the bandwidth matrix is diagonal (or constraint). This problem is discussed in detail in Sect. 5.3.3 and in Sect. 5.3.1 it is demonstrated on the basis of a simple numerical example.

Figure 5.7 shows a toy example for $M = 3$. The original vectors $\boldsymbol{c}$ and $\boldsymbol{k}$ are shown in the upper part of the figure, while the zero-padded versions are shown in the lower part. The small filled circles represent the original data and the small filled squares represent the added zeros. The resulting vectors have now the same length, making the discrete convolution theorem applicable. The following steps have to be performed:

Step 1: Compute the discrete Fourier transform for czp, i.e. C = F(czp).

Step 2: Compute the discrete Fourier transform for kzp, i.e. K = F(kzp).

Step 3: Compute the element-wise multiplication for C and K, i.e. S = CK.

Step 4: Compute the inverse Fourier transform for S, i.e. s = F-1(S).

Step 5: The desired density value is calculated as follows

$$\tilde{f}_j = P^{-1}\boldsymbol{s}\left[(2M-1):(3M-2)\right], \tag{5.25}$$

where $\boldsymbol{s}[a:b]$ stands for a subset of elements from a to b of the vector $\boldsymbol{s}$.

Note also that in (5.19) the k_l values do not have to be calculated for the full range from $l = -(M-1)$ to $l = (M-1)$, because many values are very small and can be neglected. In most practical applications, the Gaussian kernel K is used and it is true that

$$K(x) \approx 0 \quad \text{for} \quad |x| \geq \tau\sigma, \tag{5.26}$$

where σ is the standard deviation and τ is a positive number, which can be set to around 4 (the so called *effective support*). As a result, it is true that

$$K_h(\delta l) \approx 0 \quad \text{for} \quad |\delta l| \geq \tau h. \tag{5.27}$$

Here, τ was replaced by h as h has the interpretation of the standard deviation in the normal distribution, see Sect. 3.3. In that case, the summation range in (5.19) can be reduced to

$$\tilde{f}_j = \sum_{l=-L}^{L} c_{j-l} k_l, \tag{5.28}$$

where

$$L = \min\left(M-1, \left\lceil \frac{\tau h}{\delta} \right\rceil\right). \tag{5.29}$$

In practical applications, it is almost always that $L < M-1$ and, as a result, the computation burden can be significantly reduced. After applying the L-cutting defined by (5.29), Eq. (5.21) is now replaced by

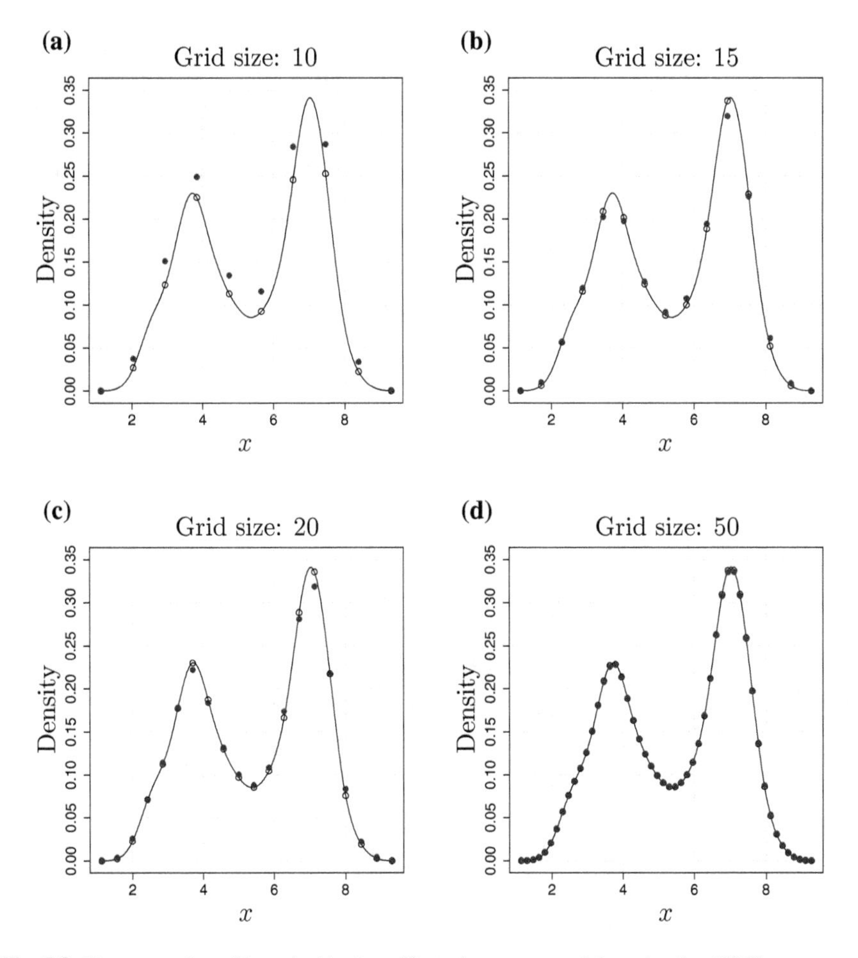

Fig. 5.8 Demonstration of how the binning affects the accuracy of the univariate KDE

$$\begin{aligned} \boldsymbol{c} &= (c_1, c_2, \ldots, c_M) \\ \boldsymbol{k} &= (k_{-L}, \ldots, k_{-1}, k_0, k_1, \ldots, k_L). \end{aligned} \tag{5.30}$$

Figure 5.8 shows how binning affects KDE of the mixture of two Gaussians (sample size $n = 200$) given by

$$f(x) = \frac{1}{2}\mathcal{N}(x;\ 4,\ 1) + \frac{1}{2}\mathcal{N}(x;\ 7,\ 0.5)\,. \tag{5.31}$$

In this example, the optimal bandwidth value is $h = 0.319$ and it was calculated using the univariate plug-in method (see Chap. 4). The solid line shows the reference KDE calculated directly from base (3.5). This reference density is compared with four different binnings with $M = \{10, 15, 20, 50\}$. $\tilde{f}_j$ was calculated using Eq. (5.10).

The individual values of densities in the grid points g_j, $j = 1, \ldots, M$ are depicted in Fig. 5.8 as the filled circles. The open circles are placed on the reference KDE precisely in the grid points g_j. For $M = 10$ (a very small and very impractical value), the binned KDE is very inaccurate (Fig. 5.8a). For $M = 15$ and $M = 20$, the errors are smaller but still easily visible (Figs. 5.8b, c). For $M = 50$, the binned KDE is almost the same as the unbinned KDE (Fig. 5.8d). In practical applications, the value of M is usually at the range of several hundreds, as suggested for example in [69]. For such big values, binned and unbinned KDEs give practically the same results, and potential errors are practically negligible.

5.3.3 Multivariate Case

The binning approximation of (3.15) is now

$$\tilde{f}(g_j, \mathbf{H}, \boldsymbol{M}) = n^{-1} \sum_{l_1=1}^{M_1} \cdots \sum_{l_d=1}^{M_d} K_{\mathbf{H}} \left(g_j - g_l \right) \boldsymbol{c}_l, \tag{5.32}$$

where g are equally spaced grid points and $\boldsymbol{c}$ are grid counts (see also examples in Sect. 5.2.2). Following the same procedure as for the univariate case, Eq. (5.32) can be rewritten so that it takes a form of the *convolution*

$$\tilde{f}_j = \sum_{l_1=-(M_1-1)}^{M_1-1} \cdots \sum_{l_d=-(M_d-1)}^{M_d-1} \boldsymbol{c}_{j-l} \boldsymbol{k}_l = \boldsymbol{c} * \boldsymbol{k}, \tag{5.33}$$

where

$$\boldsymbol{k}_l = n^{-1} K_{\mathbf{H}}(\delta_1 j_1, \ldots, \delta_d j_d), \tag{5.34}$$

and $*$ is the convolution operator.

The convolution between $\boldsymbol{c}_{j-l}$ and $\boldsymbol{k}_l$ can be effectively computed using the FFT algorithm in only $O(M_1 \log M_1 \ldots M_d \log M_d)$ operations compared to $O(M_1^2 \ldots M_d^2)$ operations required for direct computation of (5.32).

To compute the convolution between $\boldsymbol{c}$ and $\boldsymbol{k}$, these have to first be reshaped (*zero-padded*). We emphasize that the reshaping proposed in [186] is relevant only in the case when $\mathbf{H} \in \mathcal{D}$. The proposed solution presented in this book, however, supports the widest class of bandwidth matrices, that is when $\mathbf{H} \in \mathcal{F}$ (the class of symmetric, positive definite matrices) [71]. In the subsequent parts of this section, the details of the proposed improvements are presented.

For the sake of simplicity, we limit our presentation to the two-dimensional variant but the extension to higher dimensions is straightforward. The *zero-padded* $\boldsymbol{c}$ and $\boldsymbol{k}$ are as follows

$$\boldsymbol{k}_{zp} = \begin{bmatrix} \boldsymbol{k} & \boldsymbol{0} \\ \boldsymbol{0} & \boldsymbol{0} \end{bmatrix} =$$

$$\begin{bmatrix} k_{-(M_1-1),-(M_2-1)} & \cdots & k_{-(M_1-1),0} & \cdots & k_{-(M_1-1),M_2} & \\ \vdots & \ddots & \vdots & \ddots & \vdots & \\ k_{0,-(M_2-1)} & \cdots & k_{0,0} & \cdots & k_{0,M_2} & \boldsymbol{0} \\ \vdots & \ddots & \vdots & \ddots & \vdots & \\ k_{M_1-1,-(M_2-1)} & \cdots & k_{M_1-1,0} & \cdots & k_{M_1-1,M_2-1} & \cdots \\ & & \boldsymbol{0} & & \vdots & \boldsymbol{0} \end{bmatrix}, \tag{5.35}$$

and

$$\boldsymbol{c}_{zp} = \begin{bmatrix} \boldsymbol{0} & \boldsymbol{0} & \boldsymbol{0} \\ \boldsymbol{0} & \boldsymbol{c} & \boldsymbol{0} \\ \boldsymbol{0} & \boldsymbol{0} & \boldsymbol{0} \end{bmatrix} = \begin{bmatrix} \boldsymbol{0} & \vdots & \boldsymbol{0} & \vdots & \boldsymbol{0} \\ \cdots & c_{1,1} & \cdots & c_{1,M_2} & \cdots \\ \boldsymbol{0} & \vdots & \ddots & \vdots & \boldsymbol{0} \\ \cdots & c_{M_1,1} & \cdots & c_{M_1,M_2} & \cdots \\ \boldsymbol{0} & \vdots & \boldsymbol{0} & \vdots & \boldsymbol{0} \end{bmatrix}, \tag{5.36}$$

where the entry $c_{1,1}$ in (5.36) is placed in row M_1 and column M_2. Note that the proposed solution (5.35) and (5.36) are different than the ones proposed in [186]. In the proposed version of the algorithm, the problem described in Sect. 5.4 and demonstrated in Fig. 5.6 does not occur.

The sizes of the zero matrices are chosen so that after the reshaping of $\boldsymbol{c}$ and $\boldsymbol{k}$, they both have the same dimension $P_1 \times P_2, \times, \ldots, \times P_d$ (highly composite integers; typically, a power of 2). P_k $(k = 1, \ldots, d)$ are computed using to the following equation

$$P_k = 2^{\left\lceil \log_2(3M_k - 1) \right\rceil}. \tag{5.37}$$

Now, to evaluate (5.33), one can make use of the *discrete convolution theorem*, and thus perform the following operations

$$\mathbb{C} = \mathcal{F}(\boldsymbol{c}_{zp}), \quad \mathbb{K} = \mathcal{F}(\boldsymbol{k}_{zp}), \quad \boldsymbol{S} = \mathbb{C}\mathbb{K}, \quad \boldsymbol{s} = \mathcal{F}^{-1}(\mathbb{S}), \tag{5.38}$$

where $\mathcal{F}$ stands for the Fourier transform and $\mathcal{F}^{-1}$ is its inverse. The desired convolution $(\boldsymbol{c} * \boldsymbol{k})$ corresponds to a subset of $\boldsymbol{s}$ in (5.38) divided by the product of $P_1, P_2, \ldots, P_d$ (the so-called normalization), that is

$$\tilde{f}_j = (P_1\ P_2 \ldots P_d)^{-1} \boldsymbol{s}[(2M_1 - 1) : (3M_1 - 2), \ldots, (2M_d - 1) : (3M_d - 2)], \tag{5.39}$$

where, for the two-dimensional case, $\boldsymbol{s}[a:b, c:d]$ denotes a subset of rows from a to b and a subset of columns from c to d of the matrix $\boldsymbol{s}$.

In practical implementations of Wand's algorithm, the limits $\{M_1, \ldots, M_d\}$ can be additionally lowered to some smaller values $\{L_1, \ldots, L_d\}$, which significantly reduces the computational burden. In most cases, the kernel K is the multivariate normal density distribution and an *effective support* can be defined, i.e. the region outside of which the values of K are practically negligible. Now, (5.33) can be finally rewritten as

$$\tilde{f}_j = \sum_{l_1=-L_1}^{L_1} \cdots \sum_{l_d=-L_d}^{L_d} \boldsymbol{c}_{j-l} \boldsymbol{k}_l. \tag{5.40}$$

To calculate L_k $(k = 1, \ldots, d)$ the following formula is used

$$L_k = \min\left(M_k - 1, \left\lceil \frac{\tau\sqrt{|\lambda|}}{\delta_k} \right\rceil\right), \tag{5.41}$$

where λ is the largest eigenvalue of $\mathbf{H}$ and δ_k is the mesh size computed from (5.4). After a number of empirical tests, it has been discovered that τ can be set to around 3.7 for a standard two-dimensional normal kernel. Finally, the sizes P_k of matrices (5.35) and (5.36) can be calculated using to the following formula

$$P_k = 2^{\left\lceil (\log_2 (M_k + 2L_k - 1) \right\rceil}. \tag{5.42}$$

5.4 The FFT-Based Algorithm for Bandwidth Selection

5.4.1 Introductory Notes

As it was the case in Sect. 5.3.1, the original Wand's algorithm used for the task of bandwidth selection does not support unconstrained bandwidth matrices, which considerably limits its practical usefulness. We demonstrate the problems that arise in a short demonstration below (based also on the sample *Unicef* dataset).

Figure 5.9a shows the reference density when the bandwidth was obtained by direct (i.e. non-FFT-based) implementation of the LSCV algorithm presented in Sect. 4.3.2. After numerical minimization of the resulting objective function, the desired bandwidth value is derived. In Fig. 5.9b one can see the behavior of Wand's original algorithm (i.e. FFT-based) when the minimization of the objective function proceeds over $\mathbf{H} \in \mathcal{F}$. The density value is completely wrong. Figure 5.9c shows the reference density when the bandwidth was obtained by direct (i.e. non-FFT-based) implementation of the LSCV algorithm when the minimization of the objective function now proceeds over $\mathbf{H} \in \mathcal{D}$. Finally, Fig. 5.9d shows the behavior

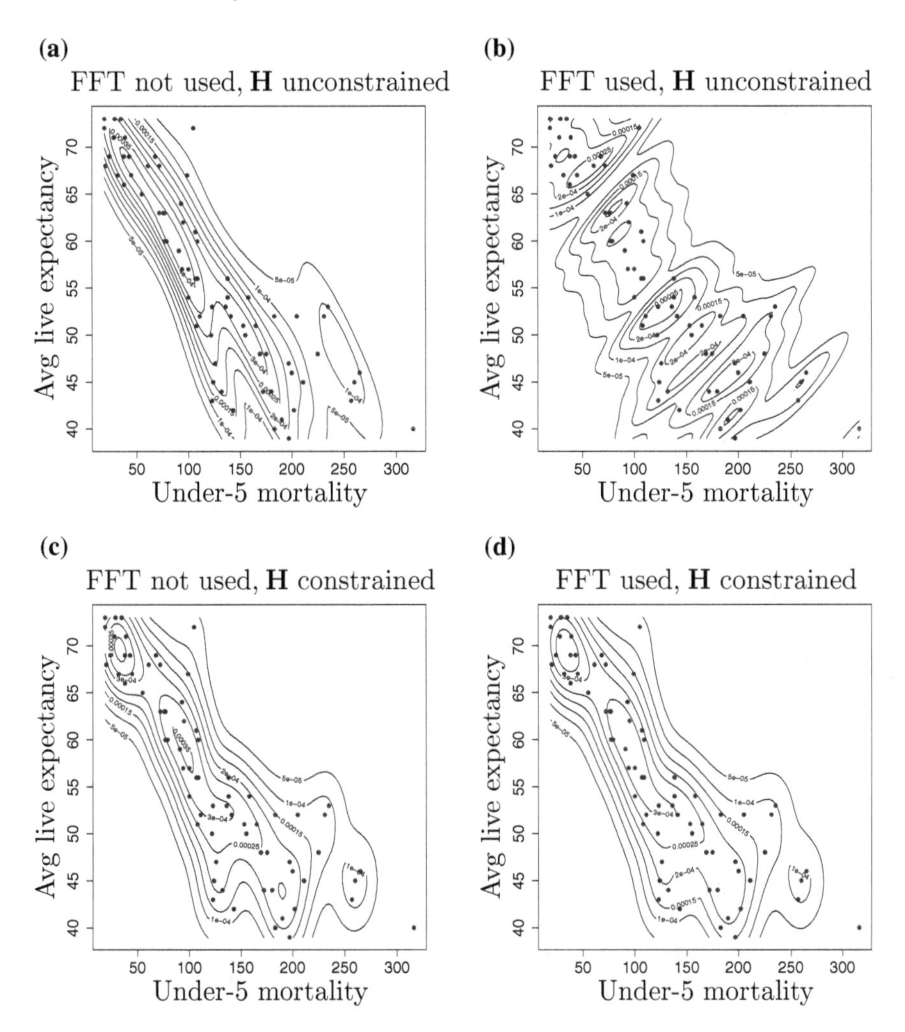

Fig. 5.9 Demonstration of behavior of Wand's original algorithm used for the task of bandwidth selection

of Wand's original algorithm when $\mathbf{H} \in \mathcal{D}$. Figures 5.9c, d are practically identical, which confirms the fact that the original version of Wand's algorithm is adequate only for constrained bandwidth matrices. Some minor differences between Figs. 5.9c, d are due to the binning of the original input data, but they are not of practical relevance.

The estimated bandwidth matrices used to plot densities shown in Figures 5.9a–d are as follows

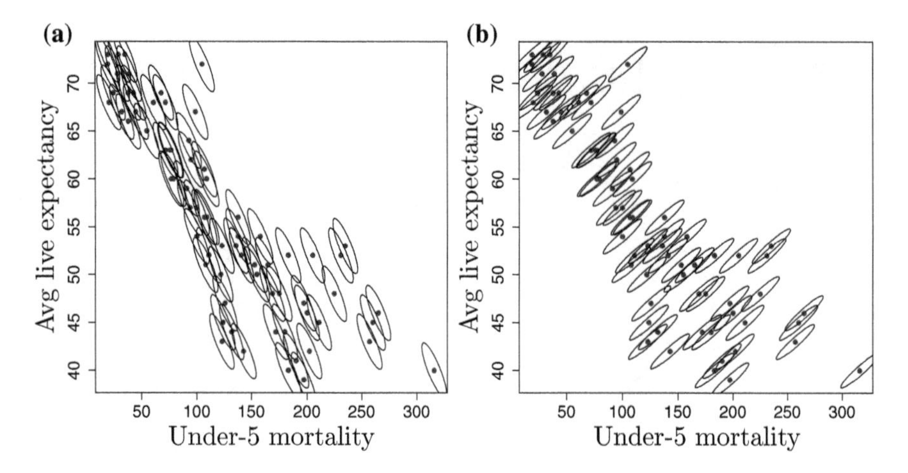

Fig. 5.10 Visualization of kernels $K_{\mathbf{H}}$ used for computing density $\hat{f}(\boldsymbol{x}, \mathbf{H})$ for the sample dataset *Unicef* (marked as small filled circles): **a** kernels generated by $\mathbf{H}_a$, **b** kernels generated by $\mathbf{H}_b$

$$\mathbf{H}_a = \begin{bmatrix} 452.34 & -93.96 \\ -93.96 & 26.66 \end{bmatrix}, \quad \mathbf{H}_b = \begin{bmatrix} 896.20 & 94.98 \\ 94.98 & 11.37 \end{bmatrix},$$
$$\mathbf{H}_c = \begin{bmatrix} 197.41 & 0.00 \\ 0.00 & 11.70 \end{bmatrix}, \quad \mathbf{H}_d = \begin{bmatrix} 242.42 & 0.00 \\ 0.00 & 11.97 \end{bmatrix}. \tag{5.43}$$

It is easy to notice that in this particular example the off-diagonal entries in $\mathbf{H}_b$ are positive, while the 'true' entries should be negative, as in $\mathbf{H}_a$. In the context of this example, it means that individual kernels $K_{\mathbf{H}}$ in (3.15) used for computing the density $\hat{f}(\boldsymbol{x}, \mathbf{H})$ are (incorrectly) 'rotated' approximately 90°, as can be seen in Fig. 5.10. The kernels generated by $\mathbf{H}_a$ bandwidth follow, correctly, the north-west dataset orientation, while $\mathbf{H}_b$ bandwidth incorrectly generates north-east oriented kernels. The $\mathbf{H}_c$ and $\mathbf{H}_d$ bandwidths do differ (due to the binning method applied) but the differences are not significant.

5.4.2 The Algorithm for the LSCV Selector

The FFT-based algorithm presented in Sect. 5.3 can also be used in the context of bandwidth selection. The main idea is exactly the same, with only some slight technical differences. To make the presentation clear, the algorithm for bandwidth selection is presented on the basis of conceptually the simplest bandwidth selection algorithm, namely LSCV, presented in Sect. 4.3.2. Here, only the multivariate case is considered, as the univariate case is just a simplified version of the former. Moreover, the univariate case for KDE was presented in a very detailed way in Sect. 5.3.2 and the extension for bandwidth selection is very natural and simple.

The starting point is the Eq. (4.34), which is repeated here for the sake of convenience

$$\mathrm{LSCV}(\mathbf{H}) = n^{-2}\sum_{i=1}^{n}\sum_{j=1}^{n} T_{\mathbf{H}}(\boldsymbol{X}_i - \boldsymbol{X}_j) + 2n^{-1}K_{\mathbf{H}}(\mathbf{0}). \tag{5.44}$$

Fast computation of the following part of (5.44) is a crucial issue

$$\hat{\psi}(\mathbf{H}) = n^{-2}\sum_{i=1}^{n}\sum_{j=1}^{n} T_{\mathbf{H}}(\boldsymbol{X}_i - \boldsymbol{X}_j). \tag{5.45}$$

This formula is in fact the estimator of the zero-th order *integrated density derivative functional*. Computational complexity of (5.45) is clearly $O(n^2)$. Thus, its fast and accurate computation plays a crucial role in bandwidth selection problems.

Coming back to the LSCV bandwidth selector, and following the same methodology as in Sect. 5.3, the input random variables $\boldsymbol{X}_j$ must be binned, so that the binned approximation of (5.45) is (5.46). Note also that now we write $\tilde{\psi}$ instead of $\hat{\psi}$ denoting this approximation.

$$\begin{aligned}\tilde{\psi}(\mathbf{H}) &= n^{-2}\sum_{i_1=1}^{M_1}\cdots\sum_{i_d=1}^{M_d}\sum_{j_1=1}^{M_1}\cdots\sum_{j_d=1}^{M_d} T_{\mathbf{H}}(g_i - g_j)\boldsymbol{c}_i\boldsymbol{c}_j \\ &= \sum_{i_1=1}^{M_1}\cdots\sum_{i_d=1}^{M_d}\boldsymbol{c}_i\left(\sum_{j_1=1}^{M_1}\cdots\sum_{j_d=1}^{M_d} T_{\mathbf{H}}(g_i - g_j)\boldsymbol{c}_j\right),\end{aligned} \tag{5.46}$$

where g are equally spaced grid points and $\boldsymbol{c}$ are grid counts and the notation is the same as that introduced in Sect. 5.2.2.

Next, the summations inside the brackets in (5.46) are rewritten so that they take a form of the convolution

$$\begin{aligned}\tilde{\psi}(\mathbf{H}) &= n^{-2}\sum_{i_1=1}^{M_1}\cdots\sum_{i_d=1}^{M_d}\boldsymbol{c}_i\left(\sum_{j_1=-(M_1-1)}^{M_1-1}\cdots\sum_{j_d=-(M_d-1)}^{M_d-1}\boldsymbol{c}_{i-j}\boldsymbol{k}_j\right) \\ &= \sum_{i_1=1}^{M_1}\cdots\sum_{i_d=1}^{M_d}\boldsymbol{c}_i\;(\boldsymbol{c} * \boldsymbol{k}),\end{aligned} \tag{5.47}$$

where

$$\begin{aligned}\boldsymbol{k}_j &= T_{\mathbf{H}}(\delta_1 j_1, \ldots, \delta_d j_d) \\ &= K_{2\mathbf{H}}(\delta_1 j_1, \ldots, \delta_d j_d) - 2K_{\mathbf{H}}(\delta_1 j_1, \ldots, \delta_d j_d).\end{aligned} \tag{5.48}$$

In the next step, the convolution between $\boldsymbol{c}_{i-j}$ and $\boldsymbol{k}_j$ can be computed using the FFT algorithm in only $O(M_1 \log M_1 \ldots M_d \log M_d)$ operations compared to the $O(M_1^2 \ldots M_d^2)$ operations required for direct computation of (5.46). To compute the convolution between $\boldsymbol{c}$ and $\boldsymbol{k}$ these have to be reshaped (*zero-padded*) first and this phase is identical as the one described in Sect. 5.3.3 (equations from (5.35) to (5.39)).

In the last step, to complete the calculations given by (5.47), the resulting d-dimensional array $(\boldsymbol{c} * \boldsymbol{k})$ needs to be multiplied by the corresponding grid counts $\boldsymbol{c}_i$ and summed to obtain $\tilde{\psi}(\mathbf{H})$, that is

$$\tilde{\psi}(\mathbf{H}) = n^{-2} \sum_{i} (\boldsymbol{c}_i \odot (\boldsymbol{c} \star \boldsymbol{k})), \tag{5.49}$$

where $\odot$ stands for the element-wise multiplication. Finally, the desired value of LSCV($\mathbf{H}$) in (5.44) can be easily and effectively calculated.

In practical implementations, the sum limits $\{M_1, \ldots, M_d\}$ can be additionally reduced to some smaller values $\{L_1, \ldots, L_d\}$, which significantly eases the computational burden (see Sect. 5.6.3 for numerical results). This phase is identical as described in Sect. 5.3.3 (equations from (5.40) to (5.42)). Now (5.47) can be rewritten as

$$\tilde{\psi}(\mathbf{H}) = n^{-2} \sum_{i_1=1}^{M_1} \cdots \sum_{i_d=1}^{M_d} \boldsymbol{c}_i \left(\sum_{j_1=-L_1}^{L_1} \cdots \sum_{j_d=-L_d}^{L_d} \boldsymbol{c}_{i-j} \boldsymbol{k}_j \right). \tag{5.50}$$

5.4.3 *Notes on the Algorithms for the PI and SCV Selectors*

From the computational point of view LSCV, PI and SCV selectors differ only in that the last two use (5.45) in a more general form. Equation (5.45) is a zero-th order *integrated density derivative functional* estimator of the general r-th order form given by

$$\hat{\psi}_r(\mathbf{H}) = n^{-2} \sum_{i=1}^{n} \sum_{j=1}^{n} \mathsf{D}^{\otimes r} K_{\mathbf{H}}(\boldsymbol{X}_i - \boldsymbol{X}_j). \tag{5.51}$$

Here, r is the derivative order (an even number), D is the gradient operator and $\mathsf{D}^{\otimes r}$ denotes a *vector* (see Sect. 4.3.4 for the explanation of the notation with $\otimes$ symbol) containing all the r-th partial derivatives of $K_{\mathbf{H}}$ arranged in the order suggested by the formal r-fold Kronecker product of the gradient operator D. This way, $\hat{\psi}_r(\mathbf{H})$ would be a *vector* of length d^r, introducing a slight complication to the FFT algorithm, especially for $d > 2$. In case of the PI and SCV bandwidth selectors, derivatives of order $r = 4, 6, 8$ are involved (see Sects. 4.2.2 and 4.3.4). Fortunately, many partial derivative values of $K_{\mathbf{H}}$ are the same and it can be proved [60, 112] that a smooth function defined on $\mathbb{R}^d$ has in general

$$\binom{d+r-1}{r} = \binom{d+r-1}{d-1}, \quad (5.52)$$

distinct partial derivatives of order r. For example for $d = 2, 3, 4, 5, 6$ and $r = 6$ the lengths of $\hat{\psi}_r(\mathbf{H})$ are

$$\{64, 729, 4096, 15625, 46656\},$$

and only

$$\{7, 28, 84, 210, 462\},$$

values are distinct. The last set of values is still rather big but not 'drastically big' and thus computer implementations can operate in acceptable times.

Note that there are two main computational problems in this context:

1. How to calculate efficiently the r -th order partial derivatives of the kernel function KH.

2. How to efficiently calculate the double sums in (5.51).

Promising algebraic solutions for the first problem have been recently developed in [22], where some efficient recursive algorithms were proposed. This paper investigates very efficient ways to compute the r-th derivatives of the multivariate Gaussian density function, that is $\mathsf{D}^{\otimes r}\phi_{\mathbf{\Sigma}}(\boldsymbol{x})$. Here

$$\phi(\boldsymbol{x}) = (2\pi)^{-d/2}\exp\left(-\frac{1}{2}\boldsymbol{x}^T\boldsymbol{x}\right), \quad (5.53)$$

is the standard d-dimensional Gaussian density and

$$\phi_{\mathbf{\Sigma}}(\boldsymbol{x}) = |\Sigma|^{-1/2}\phi(\Sigma^{-1/2}\boldsymbol{x}), \quad (5.54)$$

is the density of the multivariate normal distribution having mean zero and covariance matrix $\mathbf{\Sigma}$. All the developments in the above-mentioned paper rely on the following expression for the r-th derivative of ϕ_Σ [93]

$$\mathsf{D}^{\otimes r}\phi_{\mathbf{\Sigma}}(\boldsymbol{x}) = (-1)^r(\Sigma^{-1})^{\otimes r}\mathcal{H}_r(\boldsymbol{x}; \Sigma)\phi_{\Sigma}(\boldsymbol{x}), \quad (5.55)$$

where $\mathcal{H}_r(\boldsymbol{x}; \Sigma)$ is the r-th order Hermite polynomial.

The second problem mentioned above can be solved using the FFT-based solution presented in this chapter. In [22] a recursive algorithm for calculating the double sums in (5.51) was presented and in [72] a comparative study between this recursive algorithm and the FFT-based solution was presented. The following is generally the case: the recursive algorithms developed in [22] usually outperform the FFT-based solutions only for smaller n (around 1000–2000). For a larger n, the FFT-based implementation is much faster and, what is also very important, it is practically independent of the sample size (due to the binning used). The reader might want to consult Fig. 9 in [72] for the complete simulation results.

In a way of conclusion, we present a description of a complete procedure used for calculating (5.51) effectively, consisting of the following three steps:

Step 1: Compute all the distinct partial derivatives of order r of $K\mathbf{H}$ using a recursive algorithm developed in [22].

Step 2: Compute the double sums in (5.51) using the FFT-based algorithm presented in this chapter for the distinct partial derivatives computed in Step 1.

Step 3: Create the complete $\Psi_r(\mathbf{H})$.

The above three-steps algorithm has been already implemented in the `ks::kfe` *R* function, starting from version 1.10.0, and this implementation was based on our results presented in [71].

Note that Step 2 is definitely crucial in terms of fast computations of the final formula (5.51). In the algorithms that have been presented up until now, where methods other than the FFT-based solution proposed in this book are used, fast and accurate computations of this formula were in a sense impossible, especially for large datasets.

Note also that in the PI algorithm presented in Sect. 4.3.4, the double sums have to be computed in Steps 2, 4 and 6 and these steps are the most time consuming.

5.5 Experimental Results for Kernel Density Estimation

This section reports on experimental results of the FFT-based algorithm used for the task of KDE computation. We only describe the speed-related results, as the accuracy of the FFT-based implementation is practically identical to the case of a non-FFT-based implementation (only very small disparities can occur, typically due to issues related to the particular FFT implementation, inaccuracy of the floating point arithmetic etc.).

All the computations were performed in the *R* environment. The computations were based on artificial sample datasets of sizes $n = 100$, 500, 1000 and 2000 of a trimodal normal mixture density of the following form

$$
\begin{aligned}
f(x) &= \sum_{k=1}^{3} w_k \mathcal{N}(x;\ \mu_{k1},\ \mu_{k2},\ \sigma_{k1}^2,\ \sigma_{k2}^2,\ \delta_k) \\
&= 3/7\ \mathcal{N}\left(x;\ -2,\ -1,\ (3/5)^2,\ (7/10)^2,\ 1/4\right) \\
&+ 3/7\ \mathcal{N}\left(x;\ 1,\ 2/\sqrt{3},\ (3/5)^2,\ (7/10)^2,\ 0\right) \\
&+ 1/7\ \mathcal{N}\left(x;\ 1,\ -2/\sqrt{3},\ (3/5)^2,\ (7/10)^2,\ 0\right). \qquad (5.56)
\end{aligned}
$$

Additionally, a real dataset was used (*Unicef* dataset from the *ks R* package). The kernel K was normal and the bandwidths were calculated using the `ks::Hpi` *R* function. Grids $M = M_1 = M_2 = \{10, 20, 30, 40, 50\}$ were considered (for the purpose of making 2D plots, grid sizes greater than 50 are usually not necessary in practical applications). The following versions of *R* functions were implemented:

1. The *FFT-based* implementation (in Table 5.1 abbreviated to F). It implements Eq. (5.40). The values L_i pre-calculated according to (5.41). Grid counts $\boldsymbol{c}$ and kernels $\boldsymbol{k}$ need to be pre calculated first before the FFT convolution is performed. Computation of $\boldsymbol{k}$ is fully vectorized using typical techniques known in *R*.
2. The pure *sequential* non-FFT implementation, based on *for* loops (in Table 5.1 abbreviated to S). This version is just a literal implementation of (5.40). It takes advantage of the fact that high proportion of grid counts $\boldsymbol{c}$ are zeros (see Fig. 5.5d).
3. The maximally *vectorized* non-FFT implementation, with no *for* loops (in Table 5.1 abbreviated to V). It implements Eq. (5.32). Grid counts $\boldsymbol{c}$ and grid points g need to be pre-calculated first. This is possibly the fastest non-FFT version, as all the computations are carried out in one compact *R* command.
   It is worth to note that extensive code vectorization results in a significant growth of resources (RAM memory) needed to be allocated (if M_i is big, as 100 or more). This may be a source of rapid performance degradation, depending on actual operating system used, total RAM memory available, processor type and many other factors.

For each setting of grids ($M = M_1 = M_2$) and for each sample size n, 100 repetitions were performed and execution times were recorded. The means of the 100 repetitions constitute the final results depicted in Table 5.1.

It is easy to notice that the FFT-based implementation is absolutely unbeatable, even compared with highly vectorized codes. Vectorization gives roughly similar speedups (compared to the FFT-based method) only for very small, usually not useful in practical terms, grid sizes. Sequential implementation, unsurprisingly, is very slow and completely impractical for M_i greater than a dozen or so.

Table 5.1 Speed comparisons of the FFT-based, sequential non-FFT and vectorized non-FFT versions. The abbreviations mean as follows: F—FFT-based method, S—sequential non-FFT method, V—vectorized non-FFT method. Speedups were rounded to the nearest integer value

n	$M = M_1 = M_2$	F (sec)	V (sec)	S (sec)	V/F	S/F	S/V	L_1	L_2
100	10	0.01	0.013	0.555	1	56	43	4	4
	20	0.005	0.069	5.725	14	1145	83	7	8
	30	0.005	0.392	20.855	78	4171	53	11	12
	40	0.016	1.004	43.935	63	2746	44	14	15
	50	0.015	2.083	82.77	139	5518	40	18	19
500	10	0.017	0.021	0.28	1	16	13	2	3
	20	0.022	0.077	3.805	4	173	49	5	5
	30	0.017	0.398	14.985	23	881	38	7	7
	40	0.018	1.01	42.87	56	2382	42	9	10
	50	0.022	2.123	84.155	96	3825	40	11	12
1000	10	0.03	0.033	0.215	1	7	7	2	2
	20	0.032	0.094	2.625	3	82	28	4	4
	30	0.033	0.412	12.47	12	378	30	6	6
	40	0.032	1.011	36.16	32	1130	36	8	8
	50	0.04	2.162	72.65	54	1816	34	9	10
2000	10	0.061	0.064	0.245	1	4	4	2	2
	20	0.067	0.132	1.495	2	22	11	3	3
	30	0.07	0.455	8.73	6	125	19	5	5
	40	0.07	1.072	25.025	15	357	23	6	7
	50	0.077	2.199	56.2	29	730	26	8	8
73 (Unicef dataset)	10	0.005	0.005	0.72	1	144	144	4	9
	20	0.006	0.048	7.98	8	1330	166	8	19
	30	0.009	0.399	26.585	44	2954	67	11	29
	40	0.014	1.036	65.26	74	4661	63	15	39
	50	0.017	2.142	119.675	126	7040	56	19	49

5.6 Experimental Results for Bandwidth Selection

This section reports on experimental results of the FFT-based algorithm used for the task of bandwidth selection. The LSCV method was used in this experiment. The section is divided into three parts.

The first part reports a simulation study based on synthetic data (two-dimensional mixtures of normal densities). The advantage of using such target densities is that the exact values of the ISE can be computed between the resulting kernel density estimates and the target densities

$$\mathrm{ISE}\hat{f}(\mathbf{H}) = \int_{\mathbb{R}^d} \left(\hat{f}(\boldsymbol{x}, \mathbf{H}) - f(\mathbf{x}) \right)^2 \, d\boldsymbol{x}. \tag{5.57}$$

It was proved that the ISE of any normal mixture density has an explicit form, see for example [46].

The second part reports on a simulation study based on two real datasets. Here, the most readable way of comparing the results is to use contour plots. Given that the target densities can be more easily visualized on two-dimensional plots, we limit our presentation to two-dimensional case only.

The third part reports on the speed results in case, when the computation times needed for estimation of the optimal bandwidth matrices for both FFT-based and non-FFT-based (direct) algorithms were compared. The usability of reducing M_k into L_k is also analyzed (see (5.29) and (5.41)).

A minimization of the objective function LSCV($\mathbf{H}$) was carried out using the `stats::optim` *R* function. The *Nelder-Mead* method was used with default scaling parameters, that is the reflection factor $\alpha = 1.0$, the contraction factor $\beta = 0.5$ and the expansion factor $\gamma = 2.0$. This method was chosen as it is robust and works reasonably well for nondifferentiable functions. A disadvantage of this method is that it is relatively slow. Additionally, some numerical-like problems may also occur [72].

5.6.1 Accuracy, Synthetic Data

In this numerical experiment, the target densities shown in Fig. 4.3 were used. The sample size was $n = 1024$ and the grid sizes (for simplicity equal in each direction) were $M_1 = M_2 = \{150, 200, 250, 300\}$. For each combination of the sample size and the grid size, the ISE errors were calculated and these computations were repeated 50 times. In each repetition, a different random sample was drawn from the target density. Then, classical boxplots were drawn. Separate simulations for M_k and L_k (see ((5.47) and (5.50)) were not performed as the results are practically the same for $\tau = 3.7$ (and any bigger value). The $\hat{\mathbf{H}}_{\mathrm{LSCV}}$ bandwidth was calculated as the minimizer of the objective function (4.34).

The goal was to find answers to the following two problems: (a) whether, in general, the FFT-based algorithm gives correct results (compared with a reference textual implementation based on (4.34), where no FFT-based algorithm was used for calculating (5.45)), and (b) how the binning operation influences the final results. Figure 5.11 shows the results. Looking at the boxplots one can see that the FFT-based solution is very similar to the direct solution. Even though the ISE errors differ slightly, from the practical point of view the fluctuations can be neglected.

However, during practical experiments, certain problems of numerical nature were observed. Sometimes (mainly for the most 'fragile' model #3), after binning the data, a number of failed optimizations occurred, which manifested itself as a lack of ability to find a minimizer of the objective function LSCV(**H**) as defined in (4.33). The problem with this particular model #3 is caused by a specific data concentration, where most of the probability mass is concentrated in a very small area. Accordingly, in this case a denser griding is required. An obvious workaround of these numerical problems can be to increase the grid size.

What is also important, direct LSCV(**H**) minimization (that is without the FFT-based approach) is much more robust in the sense that there are no optimization problems as shown above. The explanation for this phenomena is that binning the data makes them highly discretized, even if there are no repeated values. This can result in a non-differentiable objective function LSCV(**H**), which is much more difficult to deal with for optimization algorithms, causing problems with finding the global minimum.

Another explanation is based on a well-known observation that the univariate LSCV criterion chooses a degenerate bandwidth $h = 0$ if too high proportion of equal observations appear in the data points. In that case, $\text{LSCV}(h) \rightarrow -\infty$ as $h \rightarrow 0$ and the cross-validation method chooses $h = 0$ as the optimal value (see [168], p. 51 for details). The issue of failure of the LSCV method for bandwidth selection in the presence of tied observations was first addressed in [30] and a method which resolves this problem was proposed in [202] (only for univariate case). A similar phenomenon could also occur in our setting, as binning produces tied observations in the transformed data (the grid points). Hence, more research is necessary to develop new or improve the existing algorithms which are more robust in terms of bandwidth estimation and are applicable in the multivariate case.

5.6.2 Accuracy, Real Data

In this section, two real datasets are analyzed. The first one is the well-known *Old Faithful Geyser Data* as investigated in [6] (and in many others). It consists of pairs of waiting times between eruptions and the durations of the eruptions for the Old Faithful geyser in Yellowstone National Park, Wyoming, USA. The dataset consists of 272 observations on 2 variables. The second dataset is the *Unicef* one available in the *ks R* package [49].

Here, the ISE criterion does not have a closed form (as opposed to any normal mixture densities), so the only sensible way to evaluate the FFT-based solution is to use contour plots. Before processing, all duplicates were discarded as all cross-validation methods are not well-behaved in this case. In the presence of duplicate observations the procedure tends to choose bandwidths that are too small. There was no need to make separate simulations for M_k and L_k (see (5.47) and (5.50)), as the results are practically the same for $\tau = 3.7$ (or any larger value). The experiment shows how the binning procedure affects the accuracy of evaluating of the objective function LSCV(**H**). Figures 5.12a, b show densities of the *Unicef* and the *Old Faithful* datasets, respectively. The optimal bandwidths were calculated based on exact solution of the objective function given by (4.34). In other words, no binning was used here. In Figs. 5.12c, d it can be observed how the binning influences the resulting densities. In this case, the calculations were based on (5.46). Note that even

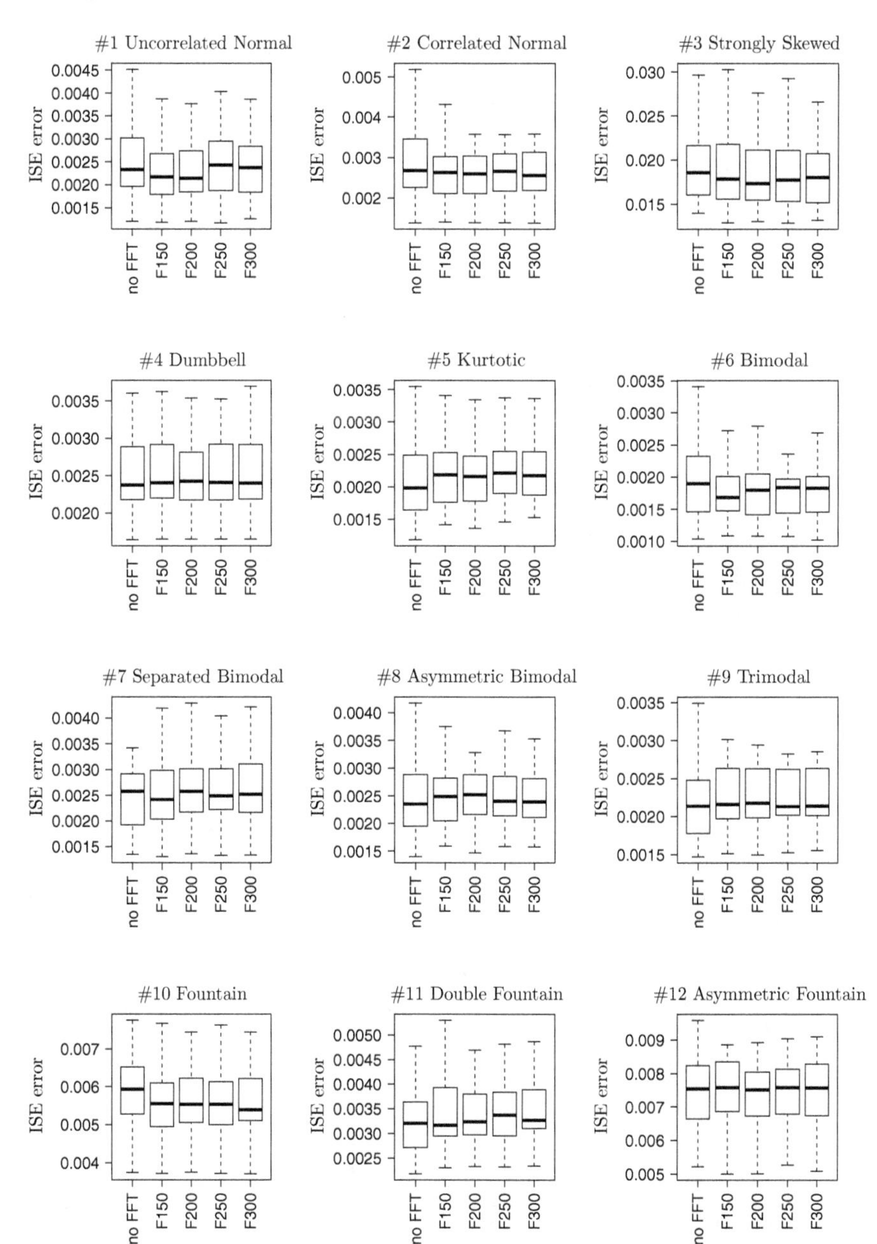

Fig. 5.11 Boxplots of the ISE errors for the sample size $n = 1024$. *no FFT* is the reference boxplot calculated without the FFT, using the direct formula (4.34). Fxx means the boxplot where gridsize xx was used

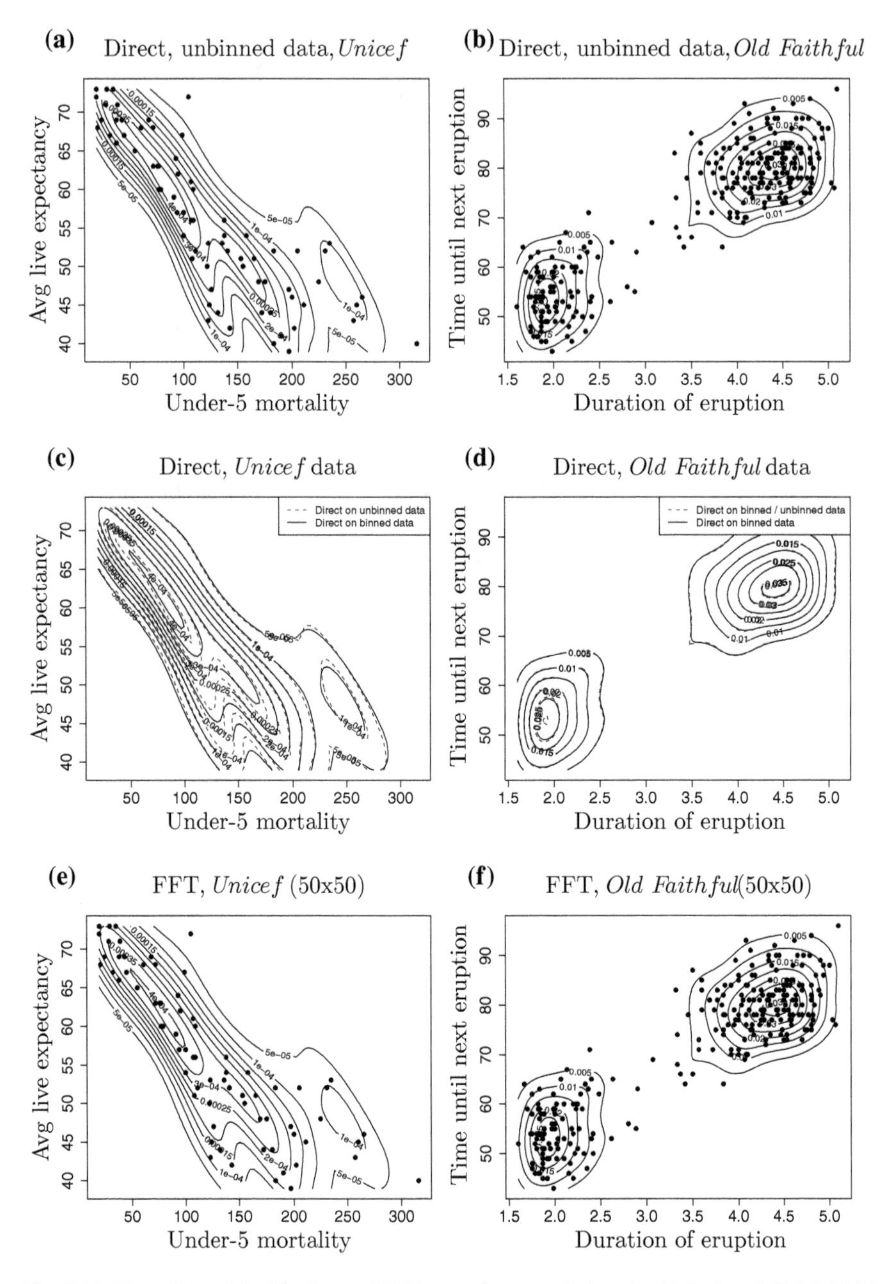

Fig. 5.12 The effect of the binning and FFT procedures applied to the *Unicef* and *Old Faithful* datasets

a moderate grid size (here $M_1 = 50$, $M_2 = 50$) is enough for our purposes and the plots differ very slightly to those presented in Figs. 5.12a, b. After making use the FFT-based approach (in this case, the calculations were based on (5.47)) the resulting contour plots presented in Figs. 5.12e, f are identical to those generated without the FFT-based support. Thus, obviously, confirming that the FFT-based procedure does find the correct bandwidths.

5.6.3 Speed Comparisons

In this section, we analyze the FFT-based approach focusing on its performance compared to non-FFT implementations. Three different *R* implementations are considered. The first one is based on direct computation of the double summations (5.45). This implementation is highly vectorized (no explicit *for* loops). The pure *for*-loops-based implementation is not analyzed here as it is extremely slow and, as such, has no practical use, especially for large n, say in the range of thousands. This implementation is called *direct*. The second implementation utilizes the FFT and is based on (5.47), where the pre-calculation of the kernel values (see (5.48)) is vectorized. This implementation is called *fft-M*. Finally, the third implementation uses (5.50), that is a modified version of (5.47), where the sum limits $\{M_1, \ldots, M_d\}$ are replaced by some smaller values $\{L_1, \ldots, L_d\}$. This implementation is called *fft-L*.

We note that this experiment does not find the minimizer $\hat{\mathbf{H}}_{LSCV}$ (4.33). This is because, in general, different runs require a different number of evaluations of the objective function under minimization. Instead, the execution times required in order to calculate functionals defined by (5.45), (5.47) and (5.50) are measured. In this experiment, the time needed for binning is also included in the results (for *fft-M* and *fft-L* methods). Binning is a necessary step and as such should not be neglected.

To reduce the number of variants, all experiments have been performed only for two-dimensional datasets. Additionally, since in this experiment the statistical structure of the dataset is not that important, the $\mathcal{N}(\mathbf{0}, \mathbf{I})$ distribution was used with the only variable being its size n. Using other distributions (i.e. these shown in Fig. 4.3) does not change the performance of the *fft-M* implementation but it can slightly affect the performance of the *fft-L* implementation. This is because different L_1 and L_2 values can be assigned (see (5.41)) and, consequently, different P_1 and P_2 values are generated (see (5.42)). Moreover, in this experiment the bandwidth $\mathbf{H}$ was fixed (its specific value is not essential here) and it was calculated using the normal reference rule (4.25).

Sample sizes from $n = 200$ to $n = 4000$, with a step size of 200, are used. Grid sizes are taken from $M_1 = M_2 = 20$ to $M_1 = M_2 = 200$ with a step size of 10 (for the sake of simplicity, grids are equal in each direction, that is $M_1 = M_2$). For the *fft-M* and *fft-L* implementations, each combination of the sample and grid sizes were used. The computations were repeated 50 times and the mean time was calculated. For the *direct* implementation, 50 repetitions were computed for each sample size and, similarly, the mean time was calculated on their basis.

Figure 5.13 shows the results for the *direct*, *fft-M* and *fft-L* implementations. One can see that for small grid sizes (roughly up to 60×60), the computation times are

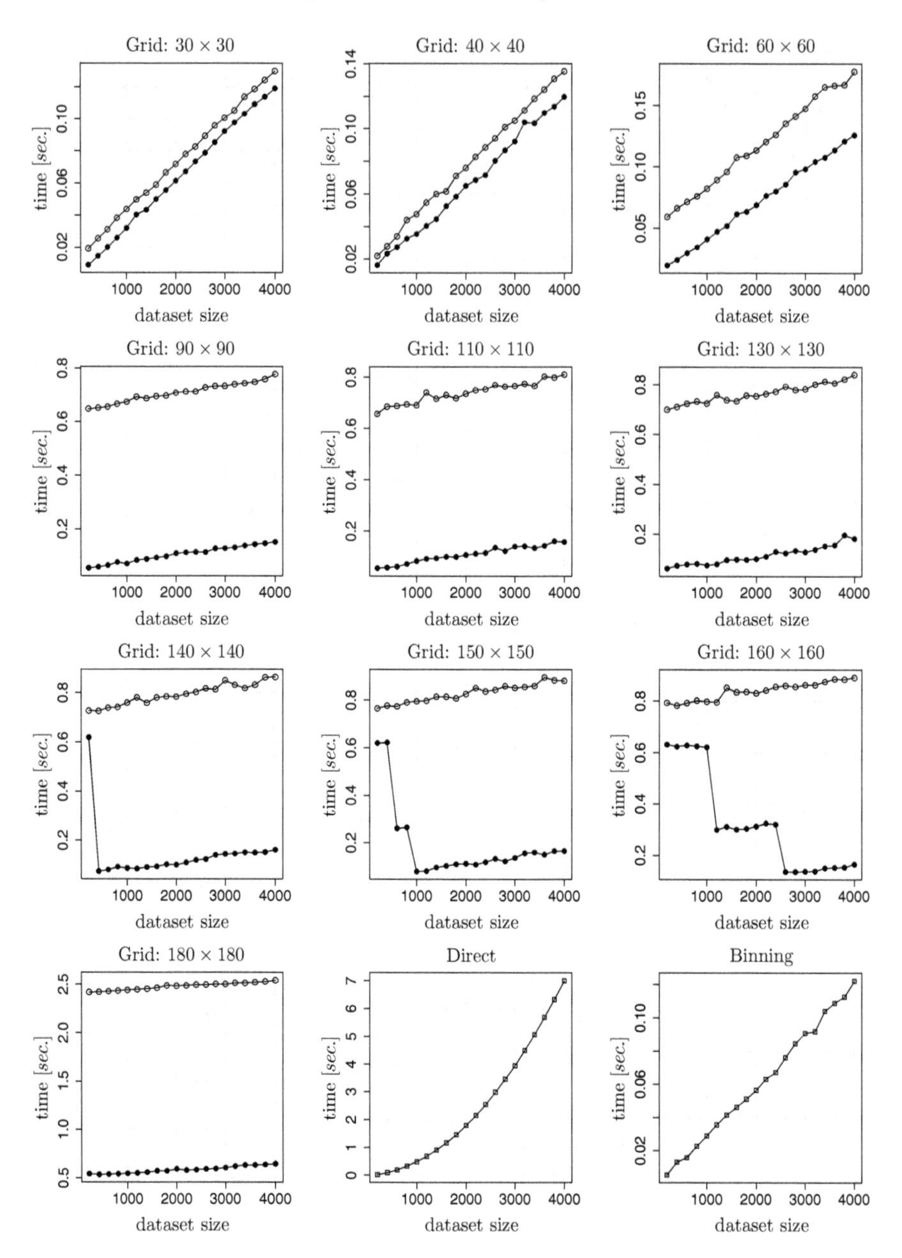

Fig. 5.13 Speed comparison results for the *fft-M*, *fft-L* and *direct* implementations. Lines marked with open circles describe the results relating to the *fft-M* implementation, lines marked with filled circles describe the results relating to the *fft-L* implementation. The most lower-right plot shows the results relating to the linear binning operation

Table 5.2 Values of P_1 and P_2 calculated using (5.42) for selected grid and sample sizes

Grid size	Sample size n									
	200	400	600	800	1000	1200	1400	1600	1800	2000
140 × 140	512	256	256	256	256	256	256	256	256	256
	512	256	256	256	256	256	256	256	256	256
150 × 150	512	512	256	256	256	256	256	256	256	256
	512	512	512	512	256	256	256	256	256	256
160 × 160	512	512	512	512	512	256	256	256	256	256
	512	512	512	512	512	512	512	512	512	512
Grid size	**Sample size n**									
	2200	2400	2600	2800	3000	3200	3400	3600	3800	4000
140 × 140	256	256	256	256	256	256	256	256	256	256
	256	256	256	256	256	256	256	256	256	256
150 × 150	256	256	256	256	256	256	256	256	256	256
	256	256	256	256	256	256	256	256	256	256
160 × 160	256	256	256	256	256	256	256	256	256	256
	512	512	256	256	256	256	256	256	256	256

more dependent on the sample size compared with the grid sizes of about 90 × 90 and bigger. Starting from the grid sizes of about 180 × 180, computation times become almost constant and this behavior is very desirable from the practical point of view. One could point out that the FFT-based implementations should not depend on n (and this is true), note however that the main portion of the computational time for smaller grid sizes comes from the binning step and not from the FFT calculations. As a consequence, a linear dependence is observed. As the binning is computed using a $O(n)$ algorithm, the observed linearity is obvious (see the most lower-right plot in Fig. 5.13).

Moreover, the *fft-L* implementation is always faster than its *fft-M* equivalent. The differences become bigger as the grid size increases. At the same time, no significant accuracy degradation is observed, so the use of L_k instead of M_k is highly recommended in practical applications.

Also, an interesting behavior in case of grid sizes 140 × 140, 150 × 150 and 160 × 160 can be observed. Namely, computation times decrease as the sample size increases. The explanation of this phenomenon is simple when one, taking into account (5.41) carefully analyses the way, in which the values of P_k were calculated. The results for the three selected grid sizes are shown in Table 5.2. For example, for the grid size 160 × 160 one can see that for sample sizes

$$n = \{200, 400, 600, 800, 1000\},$$

P_1 and P_2 are both equal to 512. For the sample sizes

$$n = \{1200, 1400, 1600, 1800, 2000, 2200, 2400\},$$

Table 5.3 Values of L_k calculated using (5.29) for some selected values of the grid and sample sizes, where $\tau = 3.7$. In parentheses, the values of L_1 and L_2 determined for given grid and sample sizes are provided

Sample size	**Grid size** ($M = M_1 = M_2$)					
	30	60	110	140	160	200
200	(14, 13)	(28, 26)	(52, 47)	(66, 60)	(75, 68)	(94, 85)
600	(10, 12)	(19, 24)	(35, 43)	(44, 55)	(51, 63)	(63, 78)
1000	(9, 11)	(19, 21)	(34, 39)	(43, 50)	(49, 57)	(61, 71)
1400	(9, 10)	(18, 20)	(33, 36)	(41, 46)	(47, 52)	(59, 65)
1800	(9, 10)	(18, 19)	(32, 35)	(41, 45)	(46, 51)	(58, 64)
2200	(9, 9)	(17, 18)	(31, 34)	(40, 43)	(45, 49)	(57, 61)
2600	(9, 9)	(17, 18)	(31, 33)	(39, 42)	(45, 48)	(56, 60)
3000	(8, 9)	(17, 18)	(30, 33)	(39, 41)	(44, 47)	(55, 59)
3400	(8, 8)	(17, 16)	(30, 29)	(38, 37)	(44, 42)	(55, 53)
3800	(8, 8)	(16, 16)	(30, 29)	(38, 37)	(43, 42)	(54, 52)

P_1 and P_2 are equal to 256 and 512, respectively. Finally, for the sample sizes

$$n = \{2600, 2800, 3000, 3200, 3400, 3600, 3800, 4000\},$$

P_1 and P_2 are both equal to 256. The values of P_k directly affect the computation time related to FFT, which causes the three 'levels' in Fig. 5.13 for the grid sizes 150×150 and 160×160.

The last two plots in Fig. 5.13 confirm the $O(n^2)$ computational complexity of the *direct* implementation and the $O(n)$ computational complexity of the binning operation. From the practical point of view, the usefulness of the *direct* implementation is very questionable, especially for large datasets.

Table 5.3 shows the values of L_k calculated using (5.41) for selected grid sizes M_k and sample sizes n. As was expected, for a given value of the grid size, its equivalents L_k are roughly the same, independently of the sample size.

5.7 Concluding Remarks

It should be stressed that the FFT-based algorithm proposed in this chapter works very fast and is very accurate as it was demonstrated by extensive computer simulations. The algorithm extends the original version on which it is based (Wand's algorithm) in such a way so that both constrained and unconstrained bandwidth matrices are supported. It was also pointed out how this algorithm can be used for fast computation of integrated density derivative functionals involving an arbitrary derivative order. This is extremely important in implementing almost all modern bandwidth selection algorithms (LSCV, SCV, PI). We note that finding a satisfactory solution for certain numerical problems reported in [72] defines an interesting and so far untackled research problem.

Chapter 6
FPGA-Based Implementation of a Bandwidth Selection Algorithm

6.1 Introduction

This chapter describes the computational aspects of bandwidth selection but it does so from the point of view of computer hardware, rather than in terms of the theory of algorithms. We present here the *Field-Programmable Gate Arrays* (FPGA) approach. The material is based mainly on the results first presented in [73].

Data binning combined with the FFT-based algorithm, as presented is Chap. 5, gives excellent results and its computational performance is also drastically improved in comparison to the other methods. For large datasets, one can reasonably argue that the FFT-based algorithm presents valuable advantages over other techniques, such as for example the one presented in [142] and extended as a technical report [143]. However, to the best of our knowledge, there is not that much research on the computational aspects related to KDE and bandwidth selection, for a review see [144]. For that reason, the FPGA-based approach presented in this chapter can be an interesting alternative to other approaches.

This chapter discusses results obtained using a tool called *Xilinx Vivado High Level Synthesis*, a feature of *Xilinx Vivado Design Suite*. This tool supports C/C++ inputs, and generates VHDL/Verilog/SystemC outputs. Other solutions are offered in *Scala* programming language [7] and a specialized high-level synthesis language called *Cx* [175]. It should also be noted that a similar tool called *A++ Compiler for HLS* is also available for Altera FPGA devices (see the web pages of these vendors for details and up-to-date versions of the mentioned tools).

The chapter is organized as follows: Sect. 6.2 provides an introduction to the designing technique known as High Level Synthesis. Section 6.3 describes a bandwidth selection method chosen to be implemented in FPGA. Section 6.4 is devoted to certain details on our FPGA implementation.

A. Gramacki, *Nonparametric Kernel Density Estimation and Its Computational Aspects*, Studies in Big Data 37,
https://doi.org/10.1007/978-3-319-71688-6_6

6.2 High Level Synthesis

Roughly speaking, there is only a number of hardware-based methods (different than using classical CPUs) that can be considered as viable options for speeding up massive computations. One can: (a) *parallelize* a given numerical algorithm and make computations in a computational cluster environment (grid computing, parallel computing) [114], (b) use dedicated hardware, i.e. *Graphics Processing Units* (GPUs) [5] (can also be configured in a grid infrastructure), (c) use modern FPGA technology [73], (d) use *Digital Signal Processors* (DSP). This chapter focuses on the third option presented above. A very gentle introduction to the FPGA technology (and its programming) can be found for example in [32, 201].

FPGA can offer significantly higher performance at much lower power costs compared with the single and multicore CPUs and GPUs in the context of many computational problems. Unfortunately, the pure programming for FPGA using *Hardware Description Languages* (HDL), such as VHDL [201] or Verilog [128] is a difficult and not-trivial task and is not intuitive for the C/C++/Java programmers. To bridge the gap between programming effectiveness and difficulty, the *High Level Synthesis* (HLS) approach is being championed by the main FPGA vendors [36, 118]. Xilinx and Altera vendors share approximately 60–90% of the entire market, these figures vary depending on the year and source. Also, many other smaller vendors offer HLS implementations, both with commercial and non-commercial (for example academic) licenses. The up-to-date list can be found for example on Wikipedia at *https://en.wikipedia.org/wiki/High-level_synthesis*.

In order to develop the FPGA design using HLS, there is no requirement for a direct HDL coding. HLS is an automated design process that interprets an algorithmic description of a problem (provided in high level languages, usually C/C++) and translates this problem into the so-called *Register-Transfer Level* (RTL) HDL code. Then, in turn, this HDL code can be easily synthesized to the gate level using logic synthesis tools, such as *Xilinx ISE Design Suite*, *Xilinx Vivado Design Suite*, *Altera Quartus II*.

Nowadays, the time-intensive computations for large datasets are mainly performed on GPU/CPU architectures but this task can also be successfully performed using the HLS approach. This chapter shows some preliminary results in terms of implementation of a selected bandwidth algorithm for KDE using HLS and presents the techniques that were used to optimize the final FPGA implementation. It also proves that FPGA speedups, compared with the highly optimized CPU and GPU implementations, are quite substantial. Moreover, the power consumption for FPGA devices is usually much lower than the typical power consumption for the present CPUs and GPUs.

It is worth to note that *OpenCL* framework, which is commonly used by GPU programmers, can also be used in the context of the FPGA devices. Nowadays, OpenCL is offered by *Altera SDK for OpenCL*, allowing one to easily implement OpenCL applications for FPGA. Recently, Xilinx announced a similar solution, namely *SDAccel Development Environment* for OpenCL, C, and C++ (see the web pages of these vendors for details).

6.3 The Method Implemented and Data Preprocessing

The direct plug-in selector presented in Sect. 4.2.2 was implemented in the FPGA architecture using the HLS approach. For convenience, we present this algorithm in a way that is ready for direct FPGA implementation. In what follows, the algorithm is abbreviated as PLUGIN (always in upper case). The individual formulas assume that the normal kernel K is used (which is a standard practice).

Input: Dataset $X_i, i = 1, \ldots n$.
Output: The optimal bandwidth h.
Step 1: Calculate the estimates of the variance $\hat{V}$ and the standard deviation $\hat{\sigma}$

$$\begin{aligned} \hat{V} &\leftarrow \frac{1}{n-1}\sum_{i=1}^{n} X_i^2 - \frac{1}{n(n-1)}\left(\sum_{i=1}^{n} X_i\right)^2, \\ \hat{\sigma} &\leftarrow \sqrt{\hat{V}}. \end{aligned} \tag{6.1}$$

Step 2: Calculate the estimate of $\Psi_8(g_8)$ using the normal scale estimate $\hat{\Psi}_8^{\mathrm{NS}}$ (formula (4.8))

$$\hat{\Psi}_8^{NS} \leftarrow \frac{105}{32\sqrt{\pi}\hat{\sigma}^9}. \tag{6.2}$$

Step 3: Calculate the bandwidth g_6 from (4.7)

$$\begin{aligned} g_6 &\leftarrow \left(\frac{-2K^6(0)}{\mu_2(K)\hat{\Psi}_8^{NS} n}\right)^{1/9}, \\ K^6(0) &= -\frac{15}{\sqrt{2\pi}}, \\ \mu_2(K) &= 1. \end{aligned} \tag{6.3}$$

Step 4: Calculate the estimate of $\Psi_6(g_6)$ using the kernel estimator $\hat{\Psi}_6(g_6)$ (see also (3.81) and (4.3))

$$\begin{aligned} \hat{\Psi}_6(g_6) &\leftarrow \frac{1}{n^2 g_6^7}\left[\sum_{i=1}^{n}\sum_{j=1}^{n} K^{(6)}\left(\frac{X_i - X_j}{g_6}\right)\right], \\ K^6(x) &= \frac{1}{\sqrt{2\pi}}\left(x^6 - 15x^4 + 45x^2 - 15\right)\exp(-\frac{1}{2}x^2). \end{aligned} \tag{6.4}$$

Step 5: Calculate the bandwidth g_4 from (4.6)

$$\begin{aligned} g_4 &\leftarrow \left(\frac{-2K^4(0)}{\mu_2(K)\hat{\Psi}_6(g_6)n} \right)^{1/7}, \\ K^4(0) &= \frac{3}{\sqrt{2\pi}}, \\ \mu_2(K) &= 1. \end{aligned} \tag{6.5}$$

Step 6: Calculate the estimate of $\Psi_4(g_4)$ using the kernel estimator $\hat{\Psi}_4(g_4)$ (see also (3.81) and (4.3))

$$\begin{aligned} \hat{\Psi}_4(g_4) &\leftarrow \frac{1}{n^2 g_4^5} \left[\sum_{i=1}^{n} \sum_{j=1}^{n} K^{(4)} \left(\frac{X_i - X_j}{g_4} \right) \right], \\ K^4(x) &= \frac{1}{\sqrt{2\pi}} \left(x^4 - 6x^2 + 3 \right) \exp(-\frac{1}{2}x^2). \end{aligned} \tag{6.6}$$

Step 7: Calculate the searched bandwidth $\hat{h}_{\mathrm{DPI},2}$ from (4.4)

$$\begin{aligned} h &\leftarrow \left(\frac{R(K)}{\mu_2(K)^2 \hat{\Psi}_4(g_4)n} \right)^{1/5} \\ R(K) &= \frac{1}{2\sqrt{\pi}}, \\ \mu_2(K) &= 1. \end{aligned} \tag{6.7}$$

It is important to stress that the PLUGIN algorithm is a strictly *sequential* computational process (see Fig. 6.1; parallel processing is possible only internally in Steps 4 and 6) as every step depends on the results obtained in the previous steps. First, the variance and the standard deviation estimators of the input data are calculated (see Step 1). Then, certain more complex formulas (in Step 2 to Step 6) are calculated. Finally, the individual values can be substitute into equation given (Step 7) yielding the desired optimal bandwidth value h.

An implementation of the algorithm above is carried out using *fixed-point arithmetic* (see Sect. 6.4.1). Unfortunately, the use of raw data while conducting the required calculations brings in a threat of potential problems in terms of overflow, especially when calculating the value of $\hat{\Psi}_8^{NS}$ (see Step 2). Note that the estimate of standard deviation in $\hat{\Psi}_8^{NS}$ is raised to the power of 9. For large values of $\hat{\sigma}$, it results in extremely small values of $\hat{\Psi}_8^{NS}$. The above problems can be successfully dealt with, if the input datasets are *standardized* using the *z-score* formula, that is

$$Z_i = \frac{X_i - \mu}{\sigma}, \tag{6.8}$$

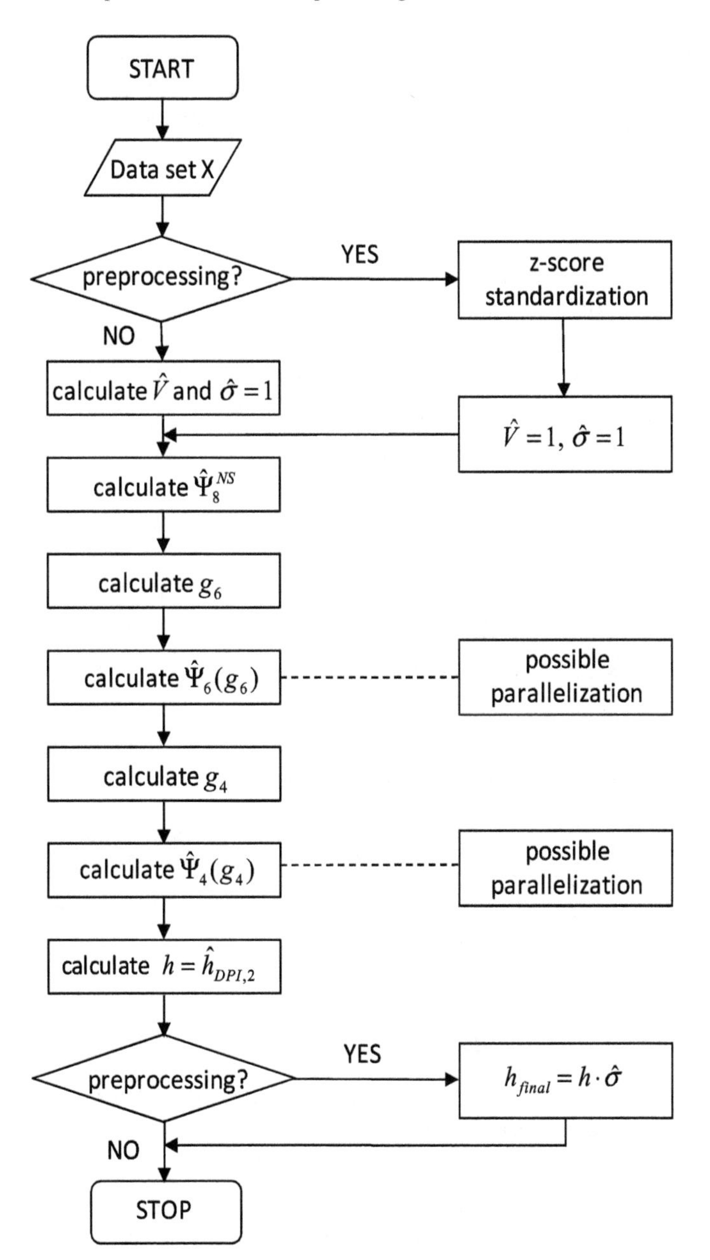

Fig. 6.1 Flowchart of the PLUGIN algorithm with optional data preprocessing (z-score standardization)

where μ and σ are mean and standard deviation of the original vector X_i, respectively. The *z-score* guarantees that $\hat{\sigma} = 1$ in $\hat{\Psi}_8^{NS}$ and, consequently, $\hat{\Psi}_8^{NS}$ entity has simply a constant value. Applying the data standardization requires preforming an extra operation on the h value in the final Step 7, that is

$$h_{\text{final}} = h \cdot \hat{\sigma}, \tag{6.9}$$

where h is the bandwidth calculated for the standardized dataset and $\hat{\sigma}$ is the standard deviation of the original vector X_i.

To reduce the computational burden, the equations for $\hat{\Psi}_6(g_6)$ and $\hat{\Psi}_4(g_4)$ can be also slightly changed. One easily notices a certain symmetry, that is

$$K^{(6)}\left(\frac{X_i - X_j}{g_6}\right) = K^{(6)}\left(\frac{X_j - X_i}{g_6}\right). \tag{6.10}$$

Therefore, the double summations can be changed and, consequently, the final formula for $\hat{\Psi}_6(g_6)$ becomes as follows:

$$\hat{\Psi}_6(g_6) \leftarrow \frac{1}{n^2 g_6^7}\left[2\left(\sum_{i=1}^{n}\sum_{j=1,i<j}^{n} K^{(6)}\left(\frac{X_i - X_j}{g_6}\right)\right) + nK^{(6)}(0)\right]. \tag{6.11}$$

Note the following details: (a) changed summation ranges, (b) the '2' factor before sums and (c) extra factor $nK^{(6)}(0)$ added. Obviously, the same concerns $K^{(4)}$ and $\hat{\Psi}_4(g_4)$, that is

$$\hat{\Psi}_4(g_4) \leftarrow \frac{1}{n^2 g_4^5}\left[2\left(\sum_{i=1}^{n}\sum_{j=1,i<j}^{n} K^{(4)}\left(\frac{X_i - X_j}{g_4}\right)\right) + nK^{(4)}(0)\right]. \tag{6.12}$$

6.4 FPGA-Based Implementation

6.4.1 Implementation Preliminaries

Before implementing the PLUGIN algorithm, it is necessary to make some assumptions that affect both the performance and resource consumption levels.

The first assumption regards the proper type of arithmetic to be used. The floating-point arithmetic provides a very good range and precision. Unfortunately, from the FPGA's point of view, this representation is very resource-demanding. In contrast, the fixed-point arithmetic is much less resource-demanding but its range and precision are more limited. Hence, the exact fixed point representation was determined based on a careful analysis of the particular intermediate values taken during calculations.

If the input dataset does not contain extremely large outliers (which suggests that such a dataset should first be carefully analyzed before any statistical analysis is performed) and if the *z-score* standardization is used, $Q32.32$ fixed point representation is sufficient for all calculations (that is: the integer part length $m = 31$, the fractional part length $n = 32$, the word length $N = 64$ and the first bit represents the sign). Also, note that as a result of the *z-score* standardization, the values of $\hat{V}$, $\hat{\sigma}$, $\hat{\Psi}_8^{NS}$ are constant and this significantly simplifies the computations.

The second assumption regards the choice of the most adequate methods for calculating individual steps of the PLUGIN algorithm. Now, it needs to be stressed that programming for FPGA devices differs considerably from programming for CPUs/GPUs devices. FPGA devices are built from a large number of simple logical blocks such as *Look Up Tables* (LUT), *Flip-Flops* (FF), *Block RAM Memory* (BRAM) and specialized DSP units. These blocks can be connected to each other and can implement only relatively low-level logical functions (referred to as the *gate level*). As a consequence, even very basic operations, such as the adder for adding two numbers, must be implemented from scratch. In the description of the PLUGIN algorithm one can easily indicate such operators like (a) addition, (b) subtraction, (c) multiplication, (d) division, (e) calculating reciprocal, (f) exponentiation, (g) taking a logarithm (note that logarithm is not explicitly used in the PLUGIN's mathematical formulas but it is used while implementing higher order roots in terms of the following definition $x^y = \exp(y \ln x)$), (h) taking power, (i) taking square roots, (j) taking higher order roots.

The implementation makes use of the following methods: *CORDIC* [184, 185] for calculating exponents and logarithms, divisions were replaced by multiplications and reciprocals, difference operators were replaced by addition of negative operands. Additionally, one extra implementation of the exponent function was used for the calculation of $K^{(6)}(x)$ and $K^{(4)}(x)$. This implementation is based on the *Remez algorithm* [38, 145] and is open to pipelining. As a consequence, a significant speedup can be achieved during calculations of Steps 4 and 6.

It is also worth to note that the authors' [73] implementation of the division operator (base on multiplications and reciprocals; the reciprocal is based on the Newton method) is significantly faster than the default division operator available in Vivado HLS. Moreover, the another advantage of using our own operators, is that no *IPCore* (Xilinx's library of many specialized functions available for FPGA projects) is needed. As a consequence, the generated VHDL codes are more portable for FPGA chips from different than Xilinx vendors.

The third assumption is related to enabling the nominal clock frequency of the FPGA chip used (see Sect. 6.4.3 for details). During experiments, it turned out that the use of the original division operator resulted in problems with obtaining the required frequency. The authors' original implementation [73] of the division operator (based on multiplications and reciprocals; the method of calculating reciprocal used the Newton method) solved this problem. For details, see the source codes in the repository [133].

The forth assumption is that all the input datasets must be stored in the BRAM memory, available in almost all current FPGA chips. They have enough capacity to store truly large data, comprising 500,000 elements or more.

6.4.2 Implementation Details

For the research purposes, we developed three different versions of the PLUGIN algorithm. The complete source codes are available for download in the PLUGIN repository [133].

The first implementation, called *literal*, is just a literal rewriting of the PLUGIN algorithm (with the improvements 6.11 and 6.12). No additional actions were taken toward optimization either in terms of the execution time or the resource requirements. This version can operate with any unscaled input data (assuming that all the inputs as well as all the internal results are represented within the fixed-point ranges that have been set). This version automatically (Vivado software is designed to make that call) utilizes pipelining. However, the pipelining does not make the implementation fast enough and, in addition one notes the use of a large number of DSP blocks. The FFs and LUTs are also very heavily made use of (see Table 6.1).

The second implementation, called *minimal*, is written so that it is optimized for resource utilization, mainly in terms of the DSP units. To reduce the number of the DSP units, dedicated functions for addition and multiplication are required. Using Vivado HLS compiler's pragmas (*#pragma HLS INLINE off*), pipelining can be disabled (on default, during translation of the high level codes into HDL, pipelining is enabled whenever it is possible). As can be observed in Table 6.1, a significant reduction of the DSP units was achieved. It confirms the fact that Vivado HLS is very sensitive in terms of the structure of the high level codes being translated into HDL ones. Therefore, in order to achieve good performance and resource usage levels, a number of high-level modifications are required.

The third implementation, called *fast*, is written so that it is optimized in terms of execution time. Addition and multiplication functions were implemented in two ways. In the first implementation (similarly as in the case of the *minimal* approach above) the pipelining is disabled, whereas in the second implementation it is enabled. The pipelined versions of the functions are used in Steps 4 and 6 of the PLUGIN algorithm as these two steps are crucial for the final performance. Additionally, in these two steps a dedicated implementation of the exponent function was used (based on Remez algorithm which is more suited to applications using pipelining). Also, a technique known as *loop unrolling* was used in a manual manner (see sample codes in Fig. 6.3).

The forth and fifth implementations (called *CPU* and *GPU* respectively) were first investigated in [5]. CPU implementation utilizes the *Streaming SIMD Extensions* (SSE) of the current multicore CPUs.

6.4.3 Results

During all experiments, the target *Xilinx Virtex-7 xc7vx690tffg1761-2* device was used. Its nominal working frequency is 200 MHz (or 5ns for a single clock tact). The CPU implementation was run on *Intel Processor i7 4790k 4.0 GHz*. The *Geforce 480GTX* graphics card was used for GPU implementation and *Vivado HLS ver. 2015.2* was used for developing all the FPGA implementations.

Table 6.1 presents a summary of the *resource consumption* together with information on *power consumption* level. The actual consumption level (in Watts) is measured by the FPGA chip after physical implementation of the PLUGIN algorithm using Vivado Design Suite. This is an estimate value and is called *Total On-Chip Power*; the power consumed internally within the FPGA, equal to the sum of *Device Static Power* and *Design Power*. It is also known as *Thermal Power*. The power consumption levels of the FPGA implementations are significantly smaller than the power consumption levels of the CPU and GPU implementations. The power consumption levels for the CPU and GPU used in the experiments turned out to be an average (catalogue-like) values.

The summary of the *execution times* for three different implementations of the PLUGIN algorithm, as well as the CPU and GPU implementations is given in Table 6.2. The *minimal* and the *fast* implementations were run on 200 MHz nominal clock while the *literal* implementation was run with 166 MHz nominal clock. This frequency degradation was caused mainly by certain limitations of the original division operator implemented in the Vivado HLS tool.

Not surprisingly, the best performance was achieved by the *fast* implementation (even compared with the *CPU* and the *GPU* implementations). This is the result of the following three optimization techniques used together: (a) an implementation of certain dedicated arithmetic operators, (b) a proper exponential function approximation and (c) the *for* loops unrolling.

A very significant speedup can be noticed, when one compares the *fast* and the *literal* implementations (average speedup about 760, see Table 6.3). The *fast* implementation is faster then the *CPU* implementation (average speedup about 32, see Table 6.3). The *fast* implementation is also faster then the *GPU* implementation (average speedup of about 10, see Table 6.3).

Table 6.1 Resource usage for three different FPGA implementations of the PLUGIN algorithms compared with the CPU and GPU implementations. Power consumption levels are also given

Method	BRAM 18k	DSP	FF	LUT	Watts
literal	128	1164	80753	81995	3.938
minimal	128	240	15889	22895	1.153
fast	128	1880	85775	38050	6.963
CPU	–	–	–	–	≈88
GPU	–	–	–	–	≈250

Table 6.2 Execution times (in sec.) for the three different FPGA implementations of the PLUGIN algorithm and for the CPU and GPU implementations

n	literal	minimal	fast	CPU	GPU
128	0.0555	0.0324	0.000276	0.0210	0.00699
256	0.2266	0.1363	0.000560	0.0252	0.00788
384	0.5155	0.3152	0.000889	0.0322	0.00947
512	0.9112	0.5513	0.001257	0.0346	0.00962
640	1.4466	0.8968	0.001667	0.0361	0.01063
768	2.1023	1.3205	0.002114	0.0375	0.01172
896	2.8771	1.8232	0.002606	0.0405	0.01447
1024	3.7666	2.3926	0.003140	0.0427	0.01641

Table 6.3 Speedups for the three different FPGA implementations of the PLUGIN algorithm and for the CPU and GPU implementations

n	literal/fast	minimal/fast	CPU/fast	GPU/fast
128	201	118	76	25
256	404	243	45	14
384	580	354	36	11
512	725	439	28	8
640	868	538	22	6
768	994	625	18	6
896	1104	700	16	6
1024	1200	762	14	5
mean value	**literal/fast**	**minimal/fast**	**CPU/fast**	**GPU/fast**
–	759.5	472.4	31.9	10.125

The summary of the *accuracy* for three different implementations of the PLUGIN algorithm is given in Table 6.4. The entity h_{ref} is the reference bandwidth calculated in double floating point arithmetic (in C++ program, 15–17 significant decimal digits) and

$$|\delta_x| = \frac{|h_{method} - h_{ref}|}{|h_{ref}|} * 100\% \tag{6.13}$$

where h_{method} means $h_{literal}$, $h_{minimal}$ or h_{fast}. It is worth noting that the relative errors for *literal*, *minimal* and *fast* implementations are very small (accounting for no more than 0.004%). In practical applications such small values can be neglected.

The summary of the *scalability* of different PLUGIN algorithm implementations is presented in Fig. 6.2. The corresponding results for *CPU* and *GPU* implementations can be found in [5]. The figure is in fact a graphical summary of data given in Table 6.2.

Table 6.4 Accuracy (relative error) for the three different FPGA implementations of the PLUGIN algorithm

n	$h_{literal}$	h_{ref}	$\|\delta_x\|$ (%)
128	0.304902711650357	0.304902701728222	3.25e-06
256	0.227651247521862	0.227651285449348	1.67e-05
384	0.202433198224753	0.202433187549741	5.27e-06
512	0.242707096505910	0.242707026022425	2.9e-05
640	0.190442902734503	0.190443702342891	0.00042
768	0.175199386896566	0.175199406819444	1.14e-05
896	0.172251554206014	0.172251524317464	1.74e-05
1024	0.174044180661440	0.174044236921001	3.23e-05
n	$h_{minimal}$	h_{ref}	$\|\delta_x\|$ (%)
128	0.304902980336919	0.304902701728222	9.14e-05
256	0.227651586290449	0.227651285449348	0.000132
384	0.202433346537873	0.202433187549741	7.85e-05
512	0.242707266006619	0.242707026022425	9.89e-05
640	0.190443017752841	0.190443702342891	0.000359
768	0.175199396442622	0.175199406819444	5.92e-06
896	0.172251742798835	0.172251524317464	0.000127
1024	0.174044403014705	0.174044236921001	9.54e-05
n	h_{fast}	h_{ref}	$\|\delta_x\|$ (%)
128	0.304901758907363	0.304902701728222	0.000309
256	0.227651913650334	0.227651285449348	0.000276
384	0.202433891594410	0.202433187549741	0.000348
512	0.242707268567756	0.242707026022425	9.99e-05
640	0.190443484811112	0.190443702342891	0.000114
768	0.175199736841023	0.175199406819444	0.000188
896	0.172251721611246	0.172251524317464	0.000115
1024	0.174044031649828	0.174044236921001	0.000118

Simplified source codes (pseudocodes) of the three FPGA implementations are presented in Fig. 6.3. Complete source codes (C++ and resulted Vivado HLS translations into VHDL) are available in the [133] repository. The first version is just the literal implementation of Step 6 in the PLUGIN algorithm written in C. Unfortunately, as can be seen in Table 6.2 and in Fig. 6.2, such implementation is very slow. In the second version, multiplications and additions are realized using certain dedicated functions (*fADD*, *fMUL*), together with a dedicated function for the reciprocal operator. In the third version, much more modifications were introduced: loop unrolling, Vivado HLS pragmas, together with multiplications and additions realized using dedicated functions with pipelining enabled (*pfADD*, *pfMUL*).

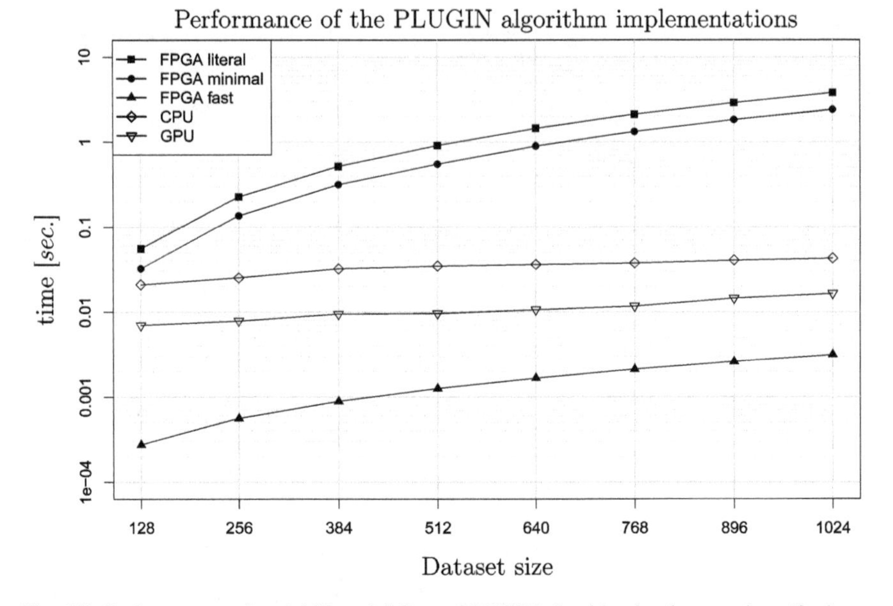

Fig. 6.2 Performance and scalability of different PLUGIN algorithm implementations (for better readability log scale for Y axis was used)

6.5 Concluding Remarks

It should be stressed that the HLS tools are competitive with manual design techniques using HDLs. The implementation time of complex numerical algorithms can be essentially reduced (compared with direct coding in HDL).

Unfortunately, to obtain efficient FPGA implementations, many changes to source code are required, compared with the equivalent implementations for CPUs and/or GPUs. This is because FPGA devices use specific primitives (DSP, BRAM, FF, LUT blocks) and programmers have to control their use in a non-automated way. However, this control is performed on the level of C/C++, and not the HDL. It is also worth emphasizing that using the HLS approach allows one to obtain the implementations that are often faster than their CPU and/or GPU counterparts.

Another crucial motivation for replacing the GPU or CPU solutions by their FPGA equivalents is the power consumption levels, which for FPGA has the value of a couple of Watts, while CPU or GPU counterparts typically take tens/hundreds of Watts or even more.

Another potential development in terms of fast implementation of numerical algorithms is related to making use of a direct HDL implementation of the PLUGIN algorithm. This is definitely much more difficult and requires much more research. However, this approach could also provide an excellent opportunity to evaluate the quality and effectiveness of the codes generated by Vivado.

```
// literal implementation
psi4_f1: for( i=0; i<N; i++ ) {
  psi4_f2: for( j=i+1; j<N; j++ ) {
    s = s + k4( ( ( x[i] - x[j] ) / g2) );
  }
}
// minimal implementation
rg2 = reciprocal( g2 );
psi4_f1: for( i=0; i<N; i++ ) {
  psi4_f2: for( j=i+1; j<N; j++ ) {
    s = fADD( s, k4( fMUL( fADD( x[i], -x[j] ), rg2 ) ) );
  }
}
// fast implementation
rg2 = reciprocal( g2 );
psi4_f1: for( i=0; i<N; i++ ) {
  psi4_f2: for( j=i+1; j<N; j+=2 ) {
    #pragma HLS EXPRESSION_BALANCE
    #pragma HLS PIPELINE
    if( j == i+1 ) tmp = 0.0;
    if( j<N ) { tmp1 = 0.0; tmp2 = 0.0; }
    psi4_f1_b0: {
      tmp1a = pfADD( x[i], -x[j] );
      tmpva = pfMUL( tmp1a, rg2 ); tmp1 = k4( tmpva );
    }
    psi4_f1_b1: {
      if( (j+1) < N ) {
        tmp1b = pfADD( x[i], -x[j+1] );
        tmpvb = pfMUL( tmp1b, rg2 ); tmp2 = k4( tmpvb );
      }
    }
    if( j<N ) {
      tmp = pfADD( tmp, tmp1 ); tmp = pfADD( tmp, tmp2 );
    }
    if( j+2>=N ) s = pfADD (s, tmp );
  }
}
```

Fig. 6.3 Three fundamental methods of the *for* loop implementation used in the $\hat{\Psi}_4(g_4)$ calculation (Step 6, Step 4 is implemented in the same way). In the *fast* implementation, the *loop unrolling* is used twice. *fADD* and *fMUL* do not use pipelining, while *pfADD* and *pfMUL* do

Last but not least, one could consider the use of modern DSP chips that offer many interesting possibilities and are potentially interesting in terms of implementing pure numerical algorithms.

Chapter 7
Selected Applications Related to Kernel Density Estimation

7.1 Introduction

The previous chapters focused primarily on the condensed presentation of the main aspects of the nonparametric kernel density estimation as a smart smoothing technique used in exploratory analysis. This chapter describes a number of selected examples of applications of this smoothing method.

It is organized as follows: Sect. 7.2 is devoted to *discriminant analysis*, a technique used to predict (or classify) a new observation to one of M known classes. This task is performed using a classifier built on a training set. The training set consists of n d-dimensional samples $\boldsymbol{X}$ grouped into M classes, where the class membership is known for every ith sample. The standard approach is to use the Bayes' Theorem in this classification task, assigning a new sample to the group with highest posterior probability. The discriminant analysis technique belongs to the group of methods known as *supervised classification*. Here, a variant of the discriminant analysis, namely *kernel discriminant analysis* is presented. Section 7.3 is devoted to *cluster analysis*, a technique used to group a set of objects in such a way that objects in the same group are (in some way) more similar to each other than to those in other groups. The cluster analysis technique belongs to the group of methods known as *unsupervised classification*. Here, a variant of the cluster analysis, namely *kernel cluster analysis* is presented. Section 7.4 shows a simple extension of the classical KDE for a *nonparametric kernel regression analysis* problem. This technique is an important complement of the well-known parametric linear regression being very popular among practitioners. Section 7.5 is devoted to a possible use of KDE methodology in *multivariate statistical process control*. This approach proves useful when the process variables differ evidently from the multivariate normal distribution. Section 7.6 describes a not-so-trivial use of the kernel smoothing method in a task of analyzing medical data coming from the *flow cytometry*.

A. Gramacki, *Nonparametric Kernel Density Estimation and Its Computational Aspects*, Studies in Big Data 37,
https://doi.org/10.1007/978-3-319-71688-6_7

7.2 Kernel Discriminant Analysis

The publications [37, 61, 105, 119] provide a short and fast-paced description of the concepts underlying discriminant analysis (DA). Let g be the number of d-dimensional independent random variables with sizes $n_1, n_2, \ldots, n_g$ drawn from g different populations, that is

$$\{\boldsymbol{X}_{11}, \ldots, \boldsymbol{X}_{1n_1}, \boldsymbol{X}_{21}, \ldots, \boldsymbol{X}_{2n_2}, \ldots, \boldsymbol{X}_{g1}, \ldots, \boldsymbol{X}_{gn_g}\} \in \mathcal{X} \tag{7.1}$$

where $\boldsymbol{X}_{ki}$ is the ith observation from the class $k \in (1, 2, \ldots, g)$. In short, the above can be written as

$$(\boldsymbol{X}_i, y_i), \quad i = 1, \ldots n, \tag{7.2}$$

where $n = n_1 + n_2 +, \ldots + n_g$ and $\boldsymbol{X}_i$ is the ith observation and y_i is the class label (labels are taken from the set $\{1, 2, \ldots, g\} \in \mathcal{G}$) to which this observation belongs.

Here, the supervised classification task is as follows: give a *decision rule* (known also as a *discriminant rule*) assigning any observation $\boldsymbol{X} \in \mathcal{X}$ to a class from a set $\mathcal{G}$, that is

$$r(\boldsymbol{x}) : \mathcal{X} \to \mathcal{G}. \tag{7.3}$$

The proper decision rule is derived on the basis of a set $(\boldsymbol{X}_i, y_i), \ i = 1, \ldots n$, known as a *learning set*. The decision rule r allows one to assign any future observation $\boldsymbol{x}$ to exactly one of the classes from the set $\mathcal{G}$.

The *kernel discriminant analysis* is one of the methods used for solving a general classification task as stated above [47, 66, 76, 123, 170] and it is based on analyzing probability density functions in classes k.

Assume that the PDF of $\boldsymbol{x}$ in class k is $p(\boldsymbol{x}|k) \equiv f_k(\boldsymbol{x})$. Now, let π_k be a priori probability that an observation belongs to the class k ($k = 1, \ldots, g$). Then, based on the Bayes' Theorem, a posteriori probability that an observation $\boldsymbol{x}$ belongs to the class k is given by

$$p(k|\boldsymbol{x}) = \frac{\pi_k \, p(\boldsymbol{x}|k)}{\sum_{i=1}^{g} \pi_i \, p(\boldsymbol{x}|i)}. \tag{7.4}$$

From the above, one is able to design the *Bayes classifier*, where a given observation is assigned to the class with the largest *posterior probability* according to the following *Bayes discriminant rule*

$$\text{Assign } \boldsymbol{x} \text{ to class } k \text{ such that } k = \underset{k \in (1,2,\ldots,g)}{\operatorname{argmax}} \ \pi_k f_k(\boldsymbol{x}). \tag{7.5}$$

If $\boldsymbol{X}_{k1}, \ldots, , \boldsymbol{X}_{kn_k}$ are d-dimensional observations in the training sample from the kth population, the kernel estimate of the density $f_k(\boldsymbol{x})$ is given by

$$\hat{f}_k(\boldsymbol{x}, \mathbf{H}_k) = n_k^{-1} \sum_{i=1}^{n_k} K_{\mathbf{H}_j}(\boldsymbol{x} - \boldsymbol{X}_{ki}). \tag{7.6}$$

The estimation of a priori probabilities π_k is easy, since their natural estimator is the sample proportion

$$\hat{\pi}_k = \frac{n_k}{n}, \tag{7.7}$$

where $n = \sum_{j=1}^{g} n_k$. The unknown class densities $f_k(\boldsymbol{x})$ are replaced by their kernel density estimates $\hat{f}_k(\boldsymbol{x}, \mathbf{H}_k)$. Finally, the *kernel discriminant analysis* (KDA) rule is as follows

$$\text{Assign } \boldsymbol{x} \text{ to class } k \text{ such that } k = \underset{k \in (1,2,\ldots,g)}{\operatorname{argmax}} \ \hat{\pi}_k \hat{f}_k(\boldsymbol{x}, \mathbf{H}_k). \tag{7.8}$$

Let us consider a mixture of three bivariate Gaussians as our example. Figure 7.1a shows the dataset and different class memberships plotted using different symbols. The univariate KDA classifier is shown in Fig. 7.1b with only the first variable x_1 taken into account. Discriminants are located on the intersections of the density curves and are marked by vertical dashed lines. Empirical densities in classes y_1, y_2, y_3 are $\hat{\pi}_1 \hat{f}_1(x_1)$, $\hat{\pi}_2 \hat{f}_2(x_2)$, $\hat{\pi}_3 \hat{f}_3(x_3)$, respectively. In this example, $n_1 = 50$, $n_2 = 150$, $n_3 = 80$ and $\hat{\pi}_1 = 0.18$, $\hat{\pi}_2 = 0.54$, $\hat{\pi}_3 = 0.28$. The classification rule is based on (7.8).

Another two toy examples use bivariate datasets with two and three classes. Figure 7.2 shows the classification results. Different class memberships are plotted using different symbols. The left column describes the first dataset, whereas the

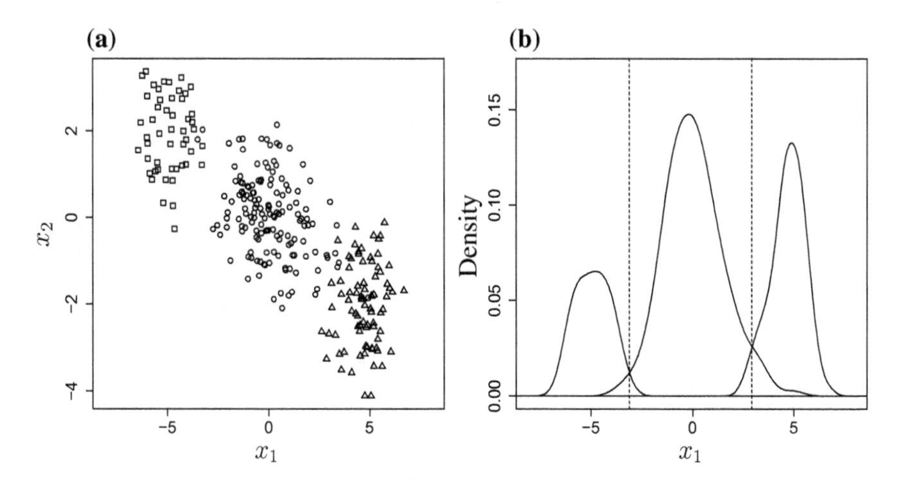

Fig. 7.1 A toy example of the univariate KDA

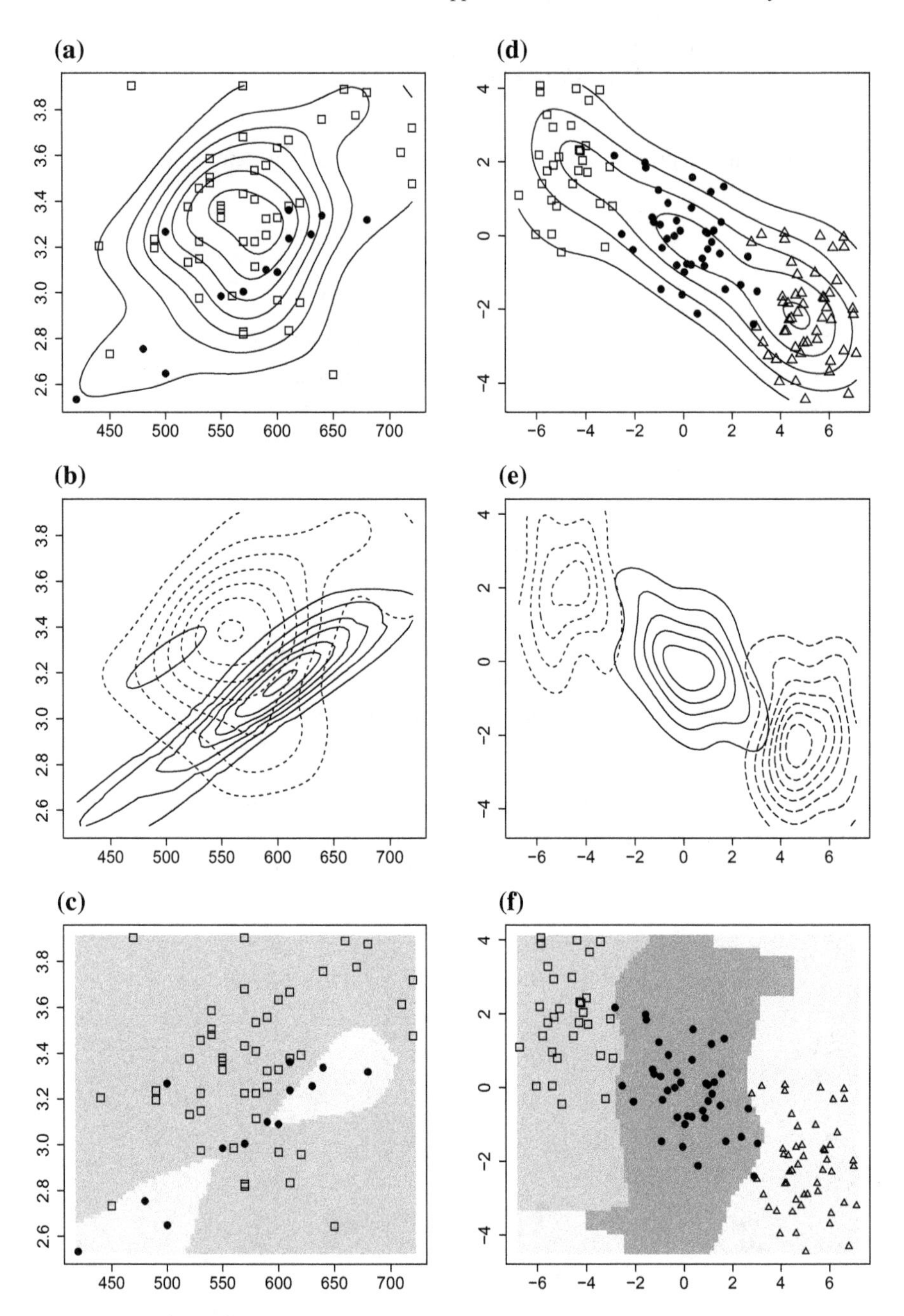

Fig. 7.2 A toy example of the bivariate KDA

right column is devoted to the second one. Figure 7.2a, d show the results of KDE of the entire datasets, and not just their fragments, without taking into account any class memberships. The multivariate plug-in selector (see Chap. 4) was used to calculate the optimal bandwidths. Figure 7.2b, e show the results of KDE in classes. For the optimal bandwidth selection, the multivariate plug-in selector was also used. The resulting KDEs take into account a priori probabilities, that is $\hat{\pi}_1 \hat{f}_1(\boldsymbol{x}_1)$, $\hat{\pi}_2 \hat{f}_2(\boldsymbol{x}_2)$ and $\hat{\pi}_3 \hat{f}_3(\boldsymbol{x}_3)$. Finally, Fig. 7.2c, f show the classification regions, marked by different gray scales. The performance of the two KDAs is very good and the individual classes are quite well separated.

7.3 Kernel Cluster Analysis

Cluster analysis (or simply clustering) is a very broad branch of statistics, for more details see for example [58]. This section focuses on the so-called *mean shift algorithm*. The publication [63] is known for introducing the mean shift algorithm. This algorithm was also presented in [168] but was in fact forgotten until it was rediscovered in [28], where the author proposed the use of the mean shift for image analysis. In another milestone paper, [34], the authors developed a complete procedure using the mean shift for the problems like image segmentation and image filtering (in that particular case it was a smoothing task). Another very interesting use of the mean shift was in terms of the real time object tracing [35]. A number of experiments related to mean shift clustering can be found in [21, 25]. An interesting problem, closely related to density derivative estimation, is that of finding filaments in point clouds. Filaments are one-dimensional curves embedded in a point process. For example, in medical imaging filaments arise as networks of blood vessels in tissue and need to be identified and mapped [64].

The mean shift algorithm is an example of *density-based clustering* [106], with DBSCAN being probably the most popular density-based clustering method [57]. Another commonly used method is the well-known *k-means* [86].

The essence of the mean shift clustering is that each object is moved to the densest area based on kernel density estimation analysis. In other words, objects converge to local maxima of density. This algorithm makes use of the density derivative as presented in Sect. 3.8. The idea is simple. Every point $\boldsymbol{X}_i$ is iteratively transformed to a new location according to the following procedure [21, 25]

$$\boldsymbol{y}_{j+1} = \boldsymbol{y}_j + \mathbf{A} \frac{\mathsf{D}\hat{f}(\boldsymbol{y}_j)}{\hat{f}(\boldsymbol{y}_j)}, \tag{7.9}$$

where $\hat{f}$ is a density estimator, $\mathsf{D}\hat{f}$ is a density gradient (1st derivative) estimator and $\mathbf{A}$ is a fixed $d \times d$ positive definite matrix chosen in such a way as to guarantee convergence of the sequences $\boldsymbol{y}_1, \boldsymbol{y}_2, \ldots$ to the local maxima of density. As can be easily seen, (7.9) differs from a well-known *gradient descent algorithm* in such a way that here the normalized density gradient is used (i.e. the density gradient is divided by the density itself). This simple trick results in a very significant acceleration of the convergence of the initial points to their densest areas, as it is be demonstrated below.

Following Sect. 3.8, the nonparametric density gradient is defined as

$$\mathsf{D}\hat{f}(\boldsymbol{x}, \mathbf{H}) = n^{-1}|\mathbf{H}|^{-1/2}(\mathbf{H}^{-1/2}) \sum_{i=1}^{n} \mathsf{D}K(\mathbf{H}^{-1/2}(\boldsymbol{x} - \boldsymbol{X}_i)). \tag{7.10}$$

Kernel K is a smooth, symmetric and unimodal function and can be expressed as

$$K(\boldsymbol{u}) = \frac{1}{2}k(\boldsymbol{u}^T\boldsymbol{u}), \tag{7.11}$$

where the function k ($k : \mathbb{R}_+ \rightarrow \mathbb{R}$) is known as the *profile* of K. As K is decreasing, then $g(x) = -k'(x) \geq 0$ and consequently

$$\mathsf{D}K(\boldsymbol{u}) = -\boldsymbol{u}g(\boldsymbol{u}^T\boldsymbol{u}). \tag{7.12}$$

Using (7.12) in (7.10) it can be shown that

$$\mathsf{D}\hat{f}(\boldsymbol{x}, \mathbf{H}) = n^{-1}|\mathbf{H}|^{-1/2}\mathbf{H}^{-1}\left(\sum_{i=1}^{n} \boldsymbol{X}_i g(\cdot) - \boldsymbol{x}\sum_{i=1}^{n} g(\cdot),\right) \tag{7.13}$$

and

$$\begin{aligned} \frac{\mathsf{D}\hat{f}(\boldsymbol{x}, \mathbf{H})}{\hat{f}(\boldsymbol{x}, \mathbf{H})} &= \frac{\mathbf{H}^{-1}\left(\sum_{i=1}^{n} \boldsymbol{X}_i g(\cdot) - \boldsymbol{x}\sum_{i=1}^{n} g(\cdot)\right)}{\sum_{i=1}^{n} g(\cdot)} \\ &= \mathbf{H}^{-1}\left(\frac{\sum_{i=1}^{n} \boldsymbol{X}_i g(\cdot)}{\sum_{i=1}^{n} g(\cdot)} - \boldsymbol{x}\right) \\ &= \mathbf{H}^{-1}\boldsymbol{m}_{\mathbf{H}}(\boldsymbol{x}), \end{aligned} \tag{7.14}$$

where

$$g(\cdot) = (\boldsymbol{x} - \boldsymbol{X}_i)^T\mathbf{H}^{-1}(\boldsymbol{x} - \boldsymbol{X}_i). \tag{7.15}$$

The term $\boldsymbol{m}_{\mathbf{H}}(\boldsymbol{x})$ is known as the *mean shift* entity. It is the difference between a weighted mean of the data and $\boldsymbol{x}$. It is also worth to note that

$$M_{\mathbf{H}}(\boldsymbol{u}, \boldsymbol{v}) = (\boldsymbol{u} - \boldsymbol{v})^T\mathbf{H}^{-1}(\boldsymbol{u} - \boldsymbol{v}), \tag{7.16}$$

is the Mahalanobis distance. Now, plugging (7.14) into (7.9) and taking $\mathbf{A} = \mathbf{H}$ leads to a recursively defined sequence

$$\boldsymbol{y}_{j+1} = \boldsymbol{y}_j + \boldsymbol{m}_{\mathbf{H}}(\boldsymbol{y}_j) = \frac{\sum_{i=1}^{n} \boldsymbol{X}_i g\left((\boldsymbol{y}_i - \boldsymbol{X}_i)^T \mathbf{H}^{-1} (\boldsymbol{y}_i - \boldsymbol{X}_i)\right)}{\sum_{i=1}^{n} g\left((\boldsymbol{y}_i - \boldsymbol{X}_i)^T \mathbf{H}^{-1} (\boldsymbol{y}_i - \boldsymbol{X}_i)\right)}. \tag{7.17}$$

Historically, the first choice, in [63], was to use a constrained bandwidth $\mathbf{H} \in \mathcal{S}$ (see Sect. 4), where $\mathcal{S} = \{h^2\mathbf{I_d} : h > 0\}$ and in [34] the authors proved that in the case of this constrained form the choice $\mathbf{A} \in \mathcal{S}$ guarantees that the mean shift sequence is convergent, as long as the kernel K has a convex and monotonically decreasing profile k. In [25] the authors proved that the unconstrained version of the mean shift algorithm (that is when $\mathbf{H} \in \mathcal{F}$) is also convergent.

In order to showcase how the mean shift clustering algorithm works, a trimodal mixture of bivariate Gaussians was used. Figure 7.3 shows the results. The three modes are marked by black big stars, the data points are marked by small black filled circles. The gray lines indicate the paths that originate from every data point and lead in the direction of a proper density mode. The small gray open circles on the paths indicate the intermediate points $\boldsymbol{y}_{j+1}$ in (7.17).

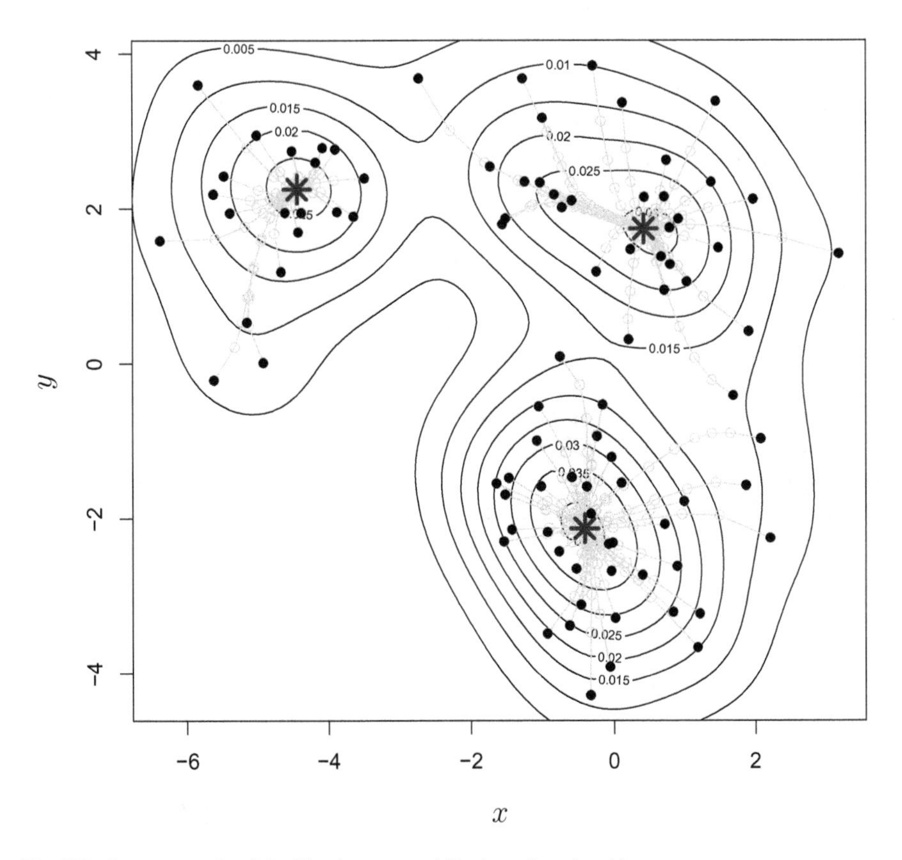

Fig. 7.3 A toy example of the bivariate mean shift clustering algorithm

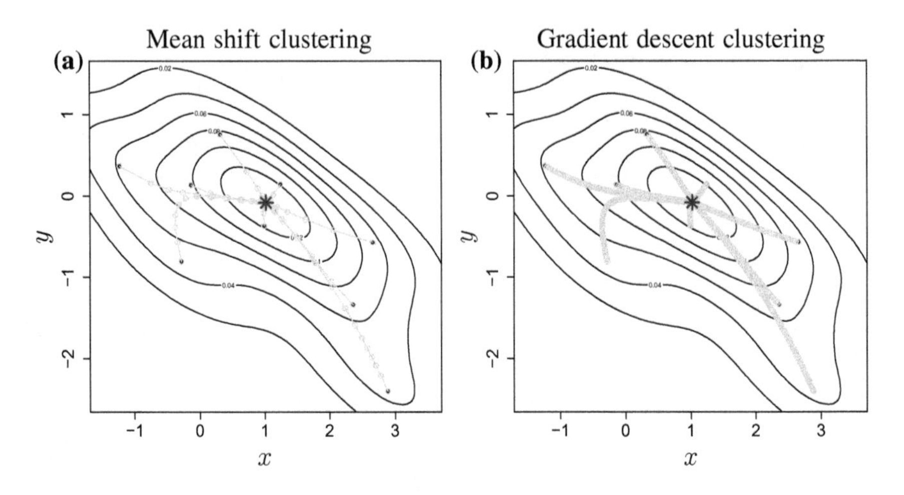

Fig. 7.4 A simple demonstration of the efficiency differences between the mean shift clustering and the gradient descent clustering

It can be easily observed that the normalization employed (the division of the density gradient by the density itself) in (7.9) has a very positive effect. Every first step from $\boldsymbol{y}_0$ to $\boldsymbol{y}_1$ is relatively long with the path determined by the point going in the right direction. To give a very positive impact of the normalization used Fig. 7.4 shows two examples (with $n = 10$). The left diagram shows the results of the mean shift clustering, while the right diagram shows the results of the use of the classical *gradient descent* algorithm, that is

$$\boldsymbol{y}_{j+1} = \boldsymbol{y}_j + \mathbf{A}\mathsf{D}\hat{f}(\boldsymbol{y}_j). \tag{7.18}$$

One easily notices that the latter is very ineffective (and consequently very slow), as many iterations are needed to approach the local extreme with a required accuracy and the individual steps are very small (or short). Moreover, the gradient descent can sometimes fail in moving a point towards a density mode if the local gradient is too small. In this example, the numbers of steps for every data point are as follows

$$\{28, 32, 36, 30, 27, 28, 31, 38, 25, 37\},$$

for mean shift clustering, and

$$\{226, 255, 308, 248, 215, 223, 255, 344, 199, 385\},$$

for gradient descent. This means that the former is about ten times faster then the latter.

7.4 Kernel Regression

This section provides a brief introduction to the *kernel nonparametric regression* (KNR), together with a simple illustrative example. Nonparametric regression has been studied by many authors with a large number of publications covering this topic, see for example [1, 13, 40, 75, 87, 138, 139, 141, 170]. Also, note that apart from KNR, a number of nonparametric smoothing techniques are available, e.g. local polynomial regression, orthogonal series smoothing and spline smoothing.

In statistics, regression analysis models the relationship between a scalar *dependent variable* Y and a number of *independent variables* (or *predictors*) X, that is

$$Y \approx m(X, \beta), \tag{7.19}$$

where β is the unknown parameter (a scalar or a vector), $(X, Y) \in \mathbb{R}^d \times \mathbb{R}$; and m is a regression function expressed as a conditional expectation function, that is

$$m(x) = \mathrm{E}(Y \mid X = x). \tag{7.20}$$

The primary goal of the regression analysis is to provide an estimate $\hat{m}$ of m from *iid* samples $(x_i, y_i), \ldots, (x_n, y_n) \in \mathbb{R}^d \times \mathbb{R}$. In nonparametric regression it is not assumed that f has any parametric form (in contrast, in very popular linear regression technique, a linear relation between Y and X is assumed). The relationship between x_i and y_i can be modeled as

$$y_i = m(x_i) + \epsilon_i, \tag{7.21}$$

where ϵ_i are *iid* random errors with zero mean and finite variance, i.e. $\mathbb{E}(\epsilon \mid X = x) = 0$ and $\mathrm{Var}(\epsilon) < \infty$.

Generally speaking, there are three different ways to model $m(x_i)$, that is:

- parametric approach,
- semiparametric approach,
- nonparametric approach.

It this section only the third approach is described in more detail.

The simplest nonparametric estimator is based on a well-known k-nearest neighbors (KNN) method, similarly to what has been described in Sect. 2.6. Here, for a fixed integer $k > 1$ the KNN regression function is defined as

$$\hat{m}_k(x) = \frac{1}{k} \sum_{i \in D_k} y_i, \tag{7.22}$$

where the summation is defined over the indexes of the k closest points of $x_1, \ldots, x_n$ to x. It is easy to notice that the regression curve $\hat{m}_k(x)$ is the mean of y_i values near a point x (this procedure is also known as *local averaging*). Varying the number of neighbors k, one achieves different estimates. The main limitation of this simple estimator is, however, that the fitted function $\hat{m}$ is characterized by unrealistic jumps (discontinuities) and the resulting curve looks jagged and has sharp edges (see Fig. 7.5a). In order to, at least partially, address this one can calculate $\hat{m}_k(x)$ using a finer grid of x but this does not eliminate jumps altogether.

Equation (7.22) can be also written as

$$\hat{m}_k(x) = \sum_{i=1}^{n} w_i(x) y_i, \tag{7.23}$$

where w_i are the weights defined as

$$w_i(x) = \begin{cases} 1/k & \text{if } x_i \text{ is one of the } k \text{ nearest points to } x \\ 0 & \text{otherwise.} \end{cases} \tag{7.24}$$

Given (7.24), it is not difficult to notice that the discontinuities of the resulted estimate $\hat{m}_k(x)$ are obviously caused by the fact that the $w_i(x)$ function is discontinuous itself.

Equation (7.22) can also be rewritten again as

$$\hat{m}_k(x) = \frac{\sum_{i=1}^{n} w_i(x) y_i}{\sum_{i=1}^{n} w_i(x)}. \tag{7.25}$$

Here, the denominator in (7.25) equals 1 for every x. Equation (7.25) is simply a classical formula describing the weighted arithmetic mean for y_i.

To avoid the problem of the discontinuities, the $w_i(x)$ function can be replaced with the smooth kernel function K, with the Gaussian or Epanechnikov kernels being a very common choice (see (3.2)). In such a case, the *Nadaraya-Watson kernel regression estimator* [124, 193] is defined as

$$\hat{m}_h(x) = \frac{\sum_{i=1}^{n} K\left(\frac{x-x_i}{h}\right) y_i}{\sum_{i=1}^{n} K\left(\frac{x-x_i}{h}\right)}. \tag{7.26}$$

Figure 7.5 shows a simple example of the nonparametric regression. This example uses the dataset *cps71*, available from the *np R* package [137] (Canadian cross-section wage data consisting of a random sample taken from the 1971 Canadian Census Public Use Tapes for male individuals having common education (grade 13). There are 205 observations in total). Figure 7.5a shows the results when the KNN estimator was used. The optimal number of nearest neighbors k was estimated by a popular rule-of-thumb method setting $k = \sqrt{n}$ (and rounding it to the nearest integer). The resulted curve is far from being smooth. Figure 7.5b shows the results when the Nadaraya-Watson estimator was used. Here, the curve is completely smooth making

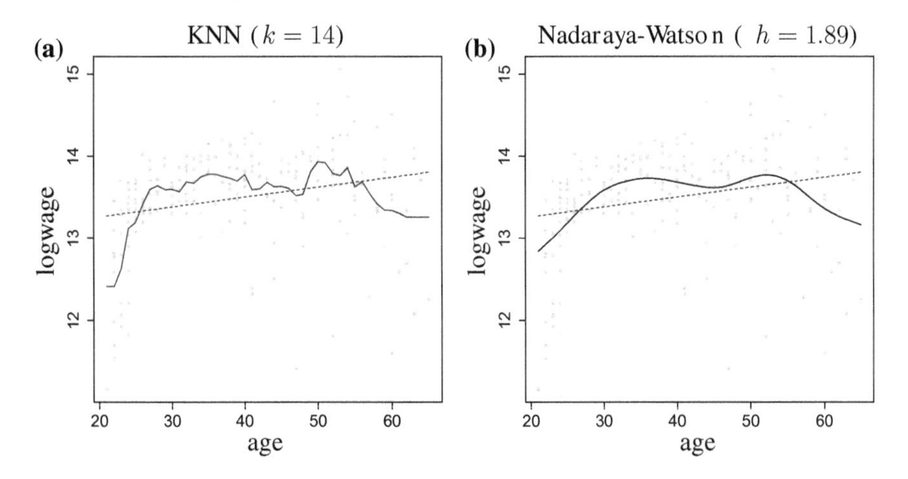

Fig. 7.5 A simple example of the nonparametric regression. (a) KNN estimator, (b) Nadaraya-Watson estimator

the method much more practical. In addition, in both plots the linear regression line is also represented (dashed).

Finally, it should be noted that the FFT-based solutions presented in Chap. 5 can obviously be used in this setting to speed up the computations performed on the basis of (7.26). The publication [140] presents another interesting FFT-based approach to the nonparametric regression estimate based on trigonometric series.

7.5 Multivariate Statistical Process Control

The statistical process control (SPC) is a very important group of methods used in industry for quality control and monitoring processes that take advantage of statistical methods. The *control charts* are the main tools used in SPC with *Shewhart control charts* [166] being probably the most common choice. The available literature on SPC is vast, with [121, 194] being some of the most often cited references.

In *univariate* SPC, only one process variable is analyzed, typically to monitor its mean $\bar{x}$, standard deviation s and range $R = x_{max} - x_{min}$. Figure 7.6 shows the three main Shewhart control chart variants generated using the *R* function `qcc::qcc` [163]. The sample dataset *pistonrings* (from the *qcc* package) was used in this example. The entities named UCL and LCL stand for *Upper Control Limit* and *Lower Control Limit*, respectively and they play a crucial role in Shewhart control charts. Values above and below these limits are often interpreted as kind of abnormal in the process under control/monitoring.

The two control limits must be estimated from the data in the so-called *Phase I* (separately for the $\bar{x}$, s and R charts). During Phase I one determines whether the

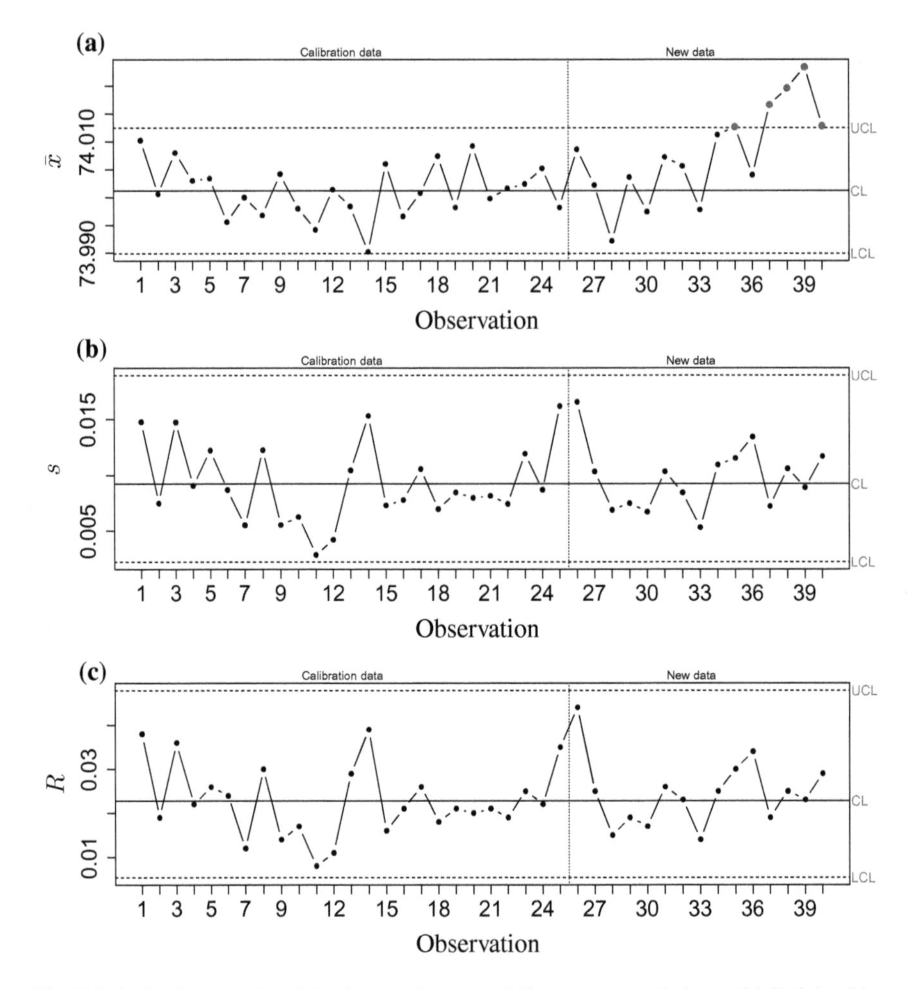

Fig. 7.6 A simple example of the three main types of Shewhart control charts: (a) $\bar{x}$ chart, (b) s chart and (c) R chart

historical data set (HDS) describes a stable (or in-control) process. In Phase II one monitors future observations, using UCL and LCL limits, and determines whether the process is still stable. In Fig. 7.6 the data points belonging to the two phases are referred to as *Calibration data* and *New data*.

In a *multivariate* SPC, the set of process variables is being analyzed as a whole and it is assumed that these variables are in a natural way correlated with each other. Hence, these should be examined together and not separately. Mason and Young [117] is the primary reference in terms of description of this research area. A multivariate SPC usually relies the *Hotelling's* T^2 statistic [95], which is a kind of generalization of the very well-known Student t statistic. The main idea of using T^2 is to reduce a d-dimensional vector of observations $\boldsymbol{x}$ to a single univariate statistic, using the following equation

$$T^2 = (\boldsymbol{x} - \boldsymbol{\mu})^T \boldsymbol{\Sigma}^{-1} (\boldsymbol{x} - \boldsymbol{\mu}). \tag{7.27}$$

The basic assumption in multivariate SPC is that the underlying d-variate process data is multivariate normal. In Phase I, the distribution of the T^2 (times a constant) can be described as the $Beta$ distribution, that is

$$T^2 = (\boldsymbol{x} - \boldsymbol{\mu})^T \boldsymbol{\Sigma}^{-1} (\boldsymbol{x} - \boldsymbol{\mu}) \sim \frac{(n-1)^2}{n} B_{(d/2,(n-d-1)/2)}. \tag{7.28}$$

In Phase II, the distribution of the T^2 (times a constant) can be described as the F distribution, that is

$$T^2 = (\boldsymbol{x} - \boldsymbol{\mu})^T \boldsymbol{\Sigma}^{-1} (\boldsymbol{x} - \boldsymbol{\mu}) \sim \frac{d(n+1)(n-1)}{n(n-d)} F_{(d,n-d)}. \tag{7.29}$$

For want of space, we omit a more detailed discussion relating the above formulas as thorough description of these can be found in the relevant references.

If the multivariate process distribution is multivariate normal, then (7.28) and (7.29) hold and the UCL of the T^2 chart is proportional to the $Beta$ or F percentiles (sometimes, to distinguish between the Phase I UCL and Phase II UCL these are written as UCL_1 and UCL_2). If, however, the multivariate process distribution is not multivariate normal, the value of such a UCL is incorrect. This may result in a raised number of false alarms. In such cases, one has to estimate the values of the T^2 statistic non-parametrically, for example using KDE methods, and determine the UCL as the $(1-\alpha)$th quantile of the fitted KDE of the T^2, where α is the false alarm rate [31].

The numerical example provided below shows this problem. To generate artificial d-variate datasets which are non-normal, we have used the *copula* approach ([126])—a great tool that is often used for modeling and simulating correlated random variables that have a difficult dependence structure. By using copulas, one can model or simulate the correlation structure and the marginals (i.e. distribution of each of the variables) of data separately. In other words, one can describe a dependence structure of data using a precisely defined expression—a copula and then 'apply' it to the marginals, which were described independently. Note that copulas provide novel application possibilities in engineering, see for example [96, 195].

There are many known copulas. Among them, the most often used are grouped into two families: *elliptical* and *Archimedean* ones. In the first family, the most popular are *normal copula* and *t-copula*. In the second, a commonly used one-parameter structures known as *Clayton*, *Frank* and *Gumbel* copulas. Their detailed description, however, goes beyond the scope of this book.

For illustrative purposes and to give the reader a feeling of how flexible the copulas are, we generate a number of 2D datasets below using five different copulas and two normal marginals ($\mu_1 = 0, \sigma_1 = 2$) and ($\mu_2 = 7, \sigma_2 = 5$), see Fig. 7.7.

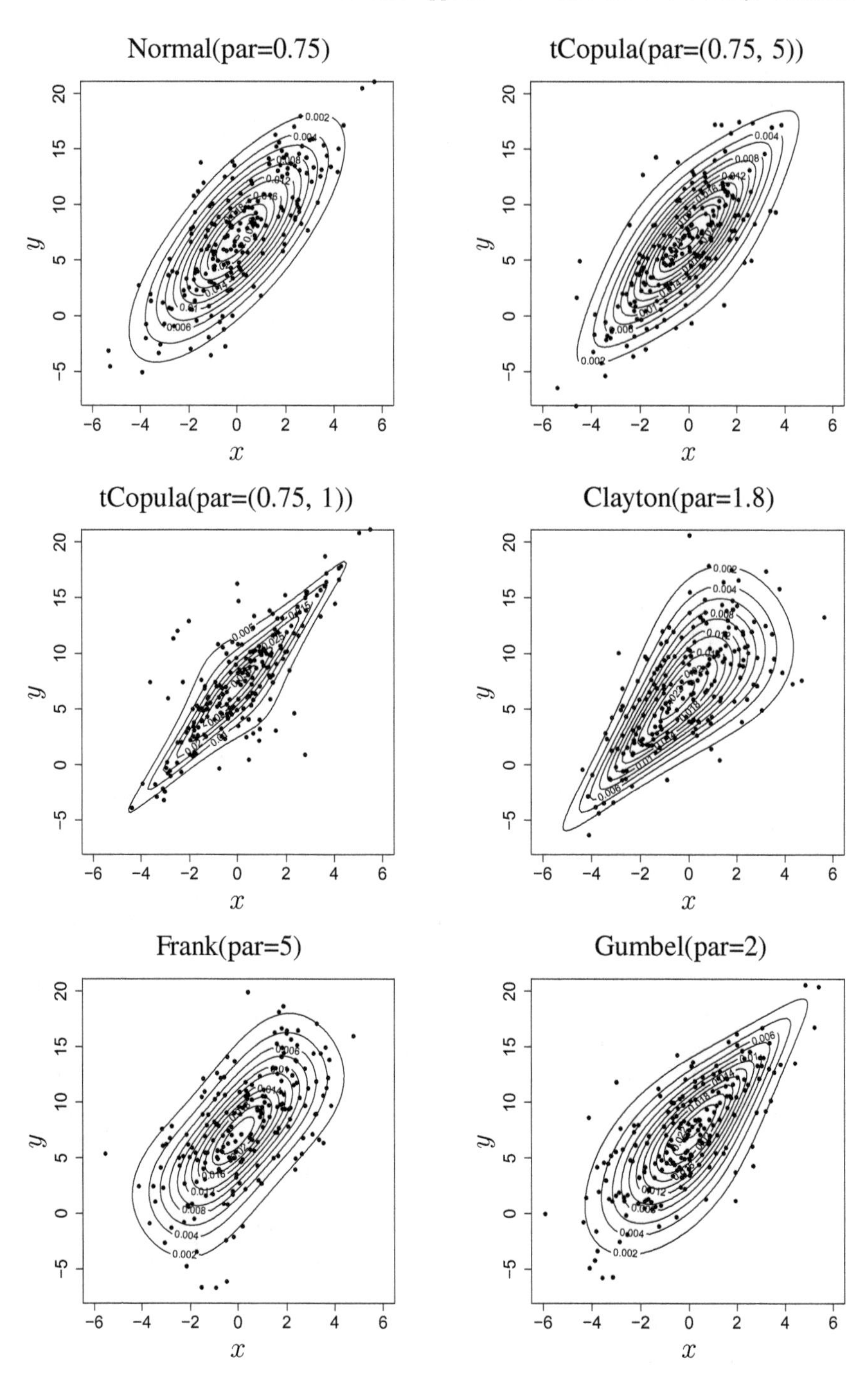

Fig. 7.7 Six example datasets generated using the copula approach

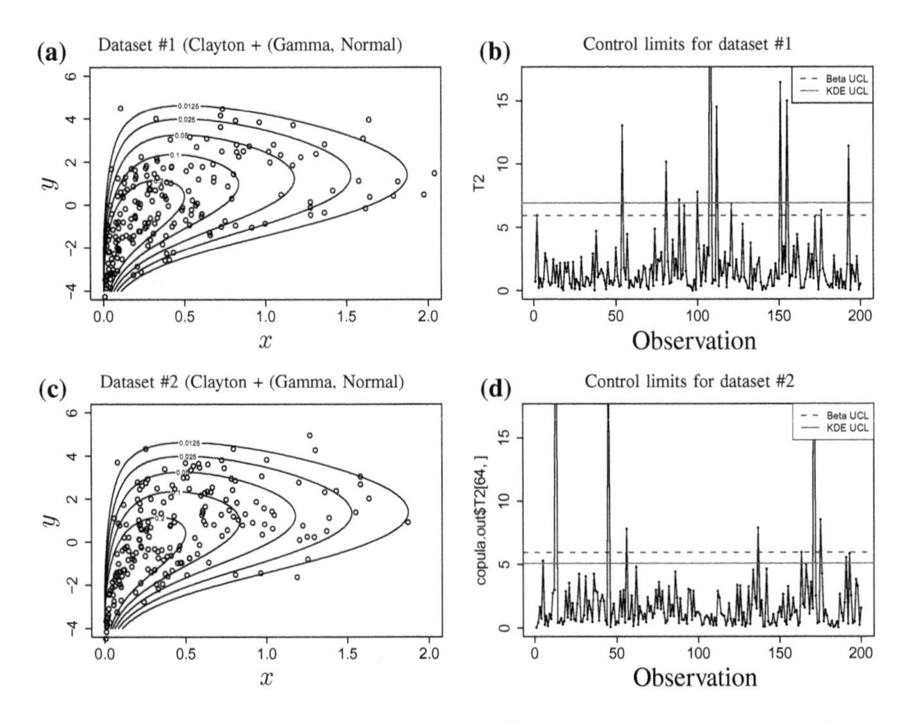

Fig. 7.8 Shewhart control charts and corresponding UCLs for two different non-normally distributed datasets (generated using the copula approach). Control limits are calculated on the basis of the *Beta* distribution and the proposed KDE-based nonparametric approach

Obviously, the marginals can be of any available form, parametric or can be even given non-parametrically. Every copula can be parametrized to tune its shape (for example 'Gumbel(2)' in Fig. 7.7 means that the Gumbel copula was used with its parameter being equal to 2). Depending on a copula type, it is parametrized using one or more parameters. Here the *copula R* package was used to: (a) specify the dependence structure of data using the copula, and (b) set marginals, in order to generate a desired random datasets.

Coming back to the SPC, the following experiment was carried out. Two different datasets have been generated using the copula approach, see Fig. 7.8a, c. The Clayton copula was used here with its parameter set to 1.8. The marginals are Gamma(shape $=$ 1, rate $=$ 2) and Normal(mean $=$ 0, std $=$ 2). Then, the classical Shewhard control chars have been prepared, see Fig. 7.8b, d.

The UCLs (UCL_1 for Phase I) were calculated in two different ways. First, by assuming that the underlying d-variate dataset is multivariate normal (see (7.28)). This UCL is displayed on the plots using a dashed line and is, of course, the same in both charts. As the datasets are not multivariate normal, this UCL, obviously, has the incorrect value and, in practical applications, this may result in incorrect number of alarms.

Therefore, to compute the UCL in a more adequate way, KDE-based approach was used (now, the UCL is determined as the $(1-\alpha)$th quantile of KDE-based fitting of the T^2). The corresponding UCLs are displayed in the plots using a solid line.

Here, an interesting behavior can be observed. In the first dataset, the UCL calculated using KDE-based approach is higher that its parametric equivalent (see Fig. 7.8b) and in the second dataset the result is opposite (see Fig. 7.8d). In that case, a natural question immediately appears: which UCL is correct? Answering this in an absolute way is not possible, as the results depend on the datasets used. It is also not difficult to generate a different dataset where the two UCLs will be roughly the same.

The above experiment is not conclusive, however. The data have been chosen in such a way that the value of UCL calculated using KDE is either above or below the value of UCL calculated in the traditional way. In order to obtain more meaningful results, one has to repeat the simulations many times and analyze the results. So, in the next experiment we have generated 500 different 2D datasets normally distributed and 500 different 2D dataset non-normally distributed (using the Clayton copula and marginals as in Fig. 7.8). Then, the corresponding UCLs have been calculated using KDE-based approach and their values have been visualized using classical boxplots. The results are shown in Fig. 7.9 and are quite interesting. Comparing the left and right boxplot, it is clear that the UCLs for non-normally distributed datasets are *higher on average* when compared to the UCLs calculated under normality assumption. This means that if the normality assumption does not hold, the classical approach (based on the F or $Beta$ distributions) should not be used as the resulted UCLs can end up being underestimated. The KDE-based approach is a much safer option in such situations.

Note that there are other KDE-based approaches described in the subject literature in the context of solving multivariate SPC problems in the presence of population's nonnormality. One should mention the bootstrap-based technique, which was used in [132]. Another very interesting result related to the nonparametric SPC has been presented in [192]. In this paper, the authors developed a framework for Shewhart charts in the space of *fuzzy* random variables. Such approach can be very attractive in situations, when it is non-numeric attributes that characterize the process under control (examples include descriptive terms such as 'good', 'bad', 'poor' etc.); or when a conversion into numeric values is not possible or at least controversial.

As the final point, it is worth noting that in many industrial applications a very important problem is the ability to monitor complex systems which generate high-dimensional data streams (hundreds and thousands of variables). Traditional methods used in SPC are restricted to lower-dimensional datasets. In [172] the author addresses this problem and proposes an interesting approach. This is done by dimensionality reduction via random projections. Consequently, even really high-dimensional process data can be effectively monitored.

7.6 Flow Cytometry

7.6.1 Introduction

This example is related to the *flow cytometry*, a method used in biophysics that is nowadays extensively used both in basic medical research and clinical diagnostics. Flow cytometry is a hi-tech laser-based technology that simultaneously measures and analyses multiple physical properties of single particles, usually human cells, as they flow in a fluid stream through a beam of laser light. The following cell features are often singled out: size, internal complexity (granularity), fluorescence intensity. Flow cytometry experiments generate huge number of multivariate data (measurements). The typical dimensionality is between 5 and 15 and the number of individual measurements vary from tens of thousands to hundreds of thousands, or even more.

A detailed explanation of flow cytometry and its practical side goes beyond the scope of this book. The available literature on this subject is really extensive, both in terms of printed books and web resources. A very gentle and condensed introduction to flow cytometry can be found in [67, 127, 136]. Readers willing looking for a more comprehensive and detailed study are invited to consult [164].

Generally speaking, every flow cytometer (a rather big, expensive and very complex technical device) generates several signals. The first two are *laser light scatter signals* when a particle (typically a human cell) deflects the laser light, see [136] and Fig. 7.10. The two signals are commonly denoted as *forward-scattered light* (FSC)

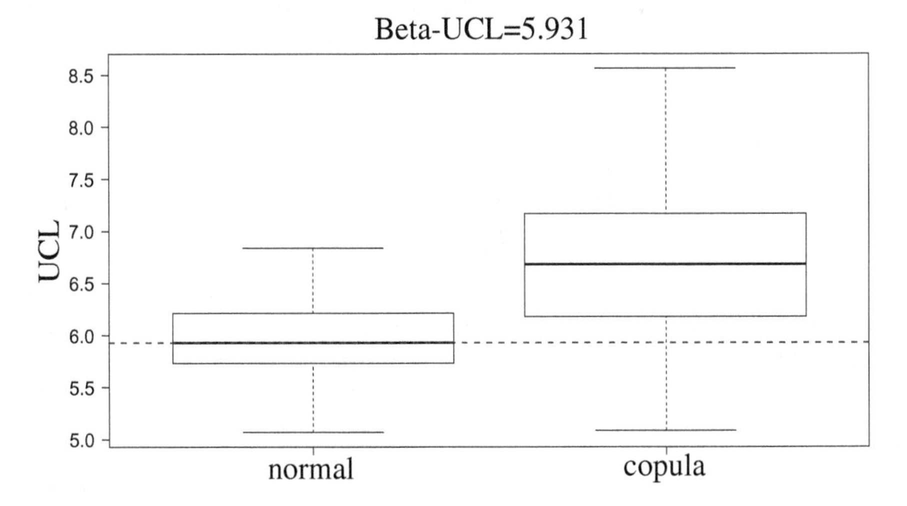

Fig. 7.9 Boxplots showing the UCLs calculated using KDE-based approach. Left boxplot: normally distributed datasets, right boxplot: non-normally distributed datasets (generated using the copula approach)

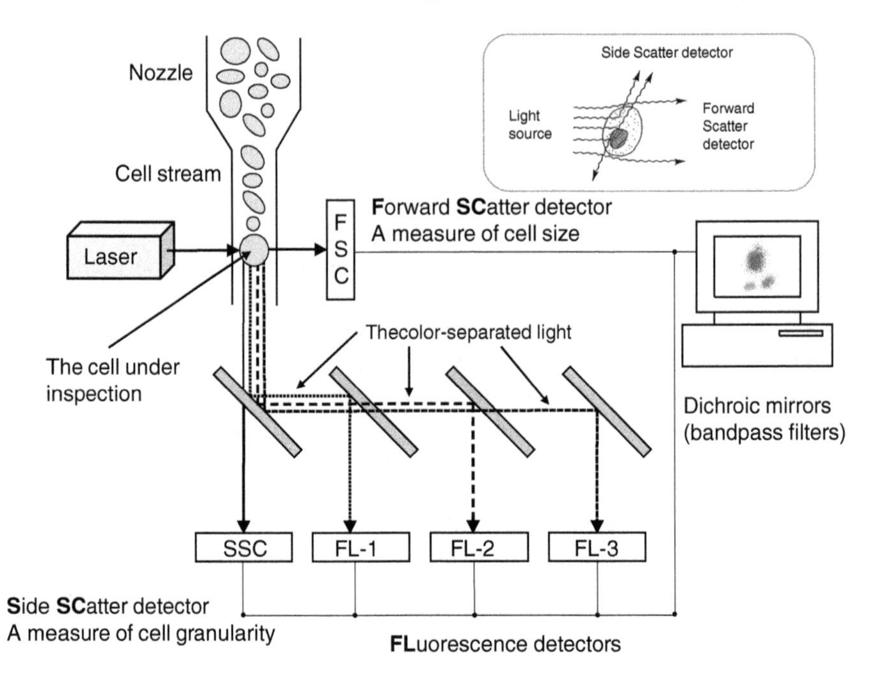

Fig. 7.10 A simplified diagram showing the main components of a typical flow cytometer device

and *side-scattered light* (SSC). The first is proportional to the cell-surface area or size. The second is proportional to the cell granularity or internal complexity. A two dimensional scatterplot of SSC and FSC allows one to differentiate among various cell types in a heterogeneous cell population, including granulocytes, monocytes, lymphocytes as well as identify other additional elements, such as debris, dead cells etc., see [136, Fig. 3.2].

The remaining signals are *fluorescence measurements*. These are a result of staining of the cells by monoclonal antibodies and are commonly denoted as *side-fluorescence light* (SFL). The signal values correspond to the intensity of the cell's fluorescence at a given laser wavelength. These signals allows one to identify many chemical and biological properties of cells.

It is also worth noting that all the flow cytometry analyses are based on statistical properties of a large number of individual cell measurements. Every single measure is in fact useless for clinical diagnosis and only a full collection of measurements can provide a valuable input. In this context, a proper data analysis is of primary significance in flow cytometry. Many analyses require one to extract (isolate, separate) cells of a particular type or characteristics. For example, in a blood sample containing a mixed population of different type cells, a researcher or doctor might want to restrict his/her analysis to lymphocytes or other subsets of cells according to physical or fluorescence measurements. This process of separating cells is called *gating*, see [136, Fig. 5.2]. Currently, gating is performed either on the basis of user's previous experience, which is obviously subjective, or on the basis of certain automatic or semi-automatic methods. Gates correspond to modal regions in the data: regions where the density function is higher that in the surrounding areas. Obviously, the notion of quality of a 'discovered' gate is difficult to define precisely. Also, it should not be surprising that two flow cytometry experts can draw 'optimal' gates in two different ways, as this process is very subjective. For that reason, the fully automatic gate drawing procedures are very difficult to implement. A good starting point in terms of the literature on gating methods is [183] and the references provided therein. In the next subsection, we describe in details a gating method closely related to KDE.

7.6.2 Feature Significance

The gating method presented in the next section relies heavily on *feature significance* for multivariate density estimation [51]. Feature significance is a method for deciding if features (such as *local maxima*, also referred to as *local modes*), are statistically significant. Another often used term in feature significance is *modal region*. A modal region corresponds to local maximum in a density function f and its surroundings. For uni- and two-dimensional data the feature significance problem was analyzed in [27, 68], respectively. In [51] the authors have extended these results for multivariate case.

As was shown in [51] finding modal regions requires testing for regions, in which the first two derivatives (i.e. gradient and curvature) in f are *significantly different from zero*. The detailed description of this method, however, goes beyond the scope of this book and interested reader should consult the relevant references.

Figure 7.11 shows a bivariate toy example of the feature significance identification. We made use of the dataset *geyser* from the *MASS R* package (the eruptions data from the 'Old Faithful' geyser in Yellowstone National Park, Wyoming, USA) with 299 observations with 2 variables. By the way, note that there are a few versions of this dataset available on the Internet. For example, in the previous sections, the dataset *faithful* from the *datasets R* package was used. The dataset used in this example (*geyser*) is very similar to the *faithful* one but its dataset is slightly smaller.

The significant curvature regions are marked on the plot as three polygons. The significant points that fall inside these regions are plotted as black filled points. The points that are outside the three regions are plotted in gray. Feature significance methods are implemented in the *feature R* package [55].

7.6.3 A Gating Method

The feature significance technology briefly presented in the previous section has been used in developing a complete framework for gating flow cytometry [125].

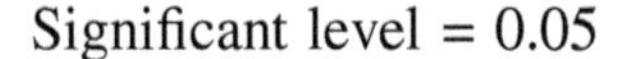

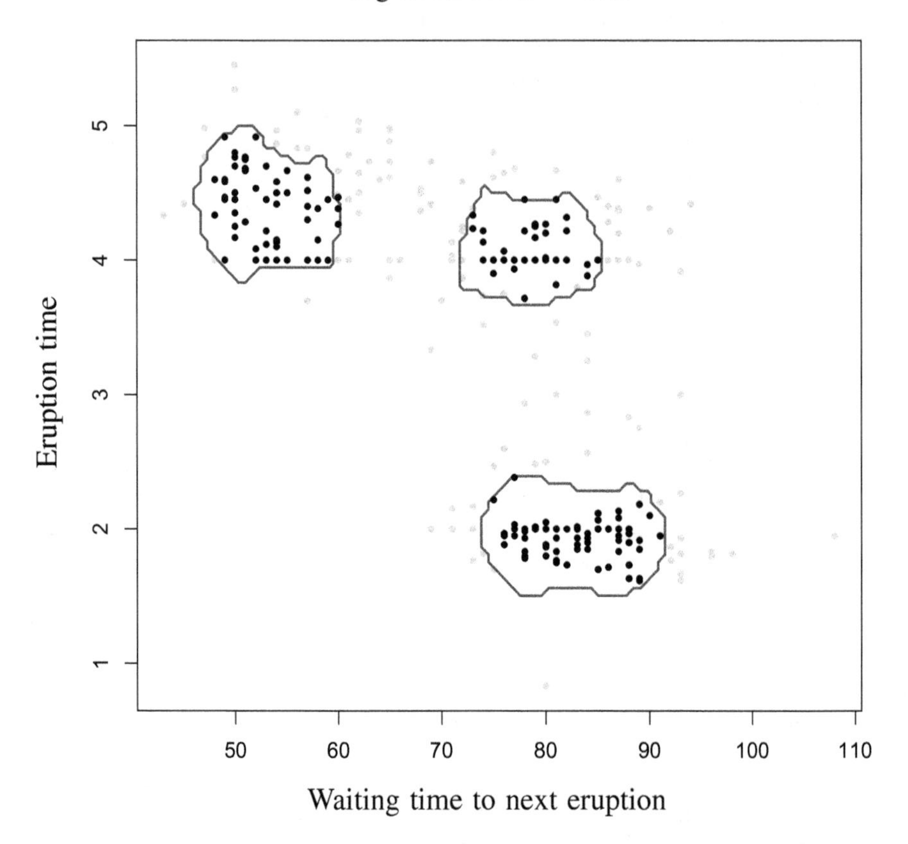

Fig. 7.11 A toy example of the bivariate feature significance determination of the curvature of a sample kernel density estimate

The framework was implemented in the *curvHDR R* package [189]. A sample flow cytometry dataset used in this presentation was described in [70] and can be downloaded from *Bioconductor* portal—the open source software for bioinformatics [10] (use the link given in the reference provided in [70]).

This dataset consists of 14 patient samples from 2 different groups treated with either drug A or B. Each sample has been stained for CD3, CD4, CD8, CD69 and HLADR. The data are stored in the flow cytometry data file standard *FCS 3.0* and can be read into *R* using *flowCore* package available at the Bioconductor portal. Note that this is one of a number of packages used for working with flow cytometry that are made available at the portal. These packages are not available in the official R packages repository [135]) and must be downloaded directly from the Bioconductor portal and installed using its specific installation tool.

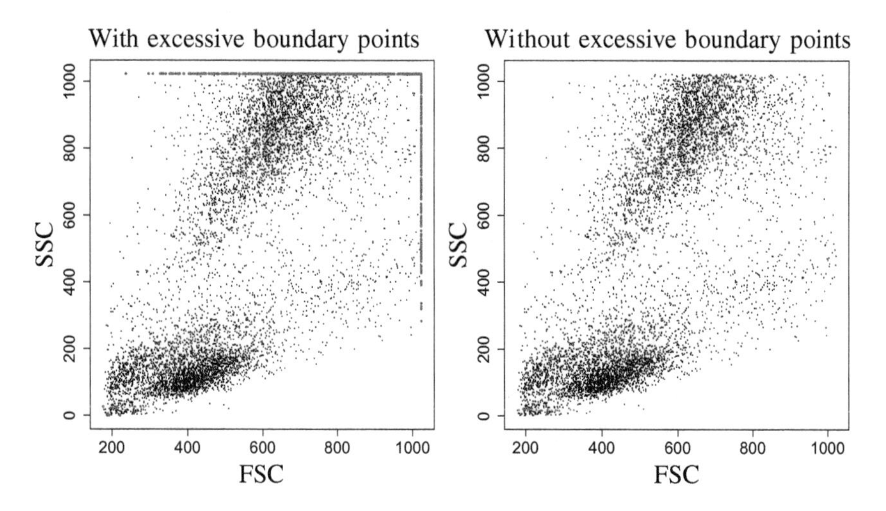

Fig. 7.12 A sample flow cytometry dataset with and without excessive boundary points

The following steps form a complete framework leading to gating flow cytometry data.

Step 1: Clear the data, e.g. remove cell debris (dead cells), remove evident outliers, excessive boundary points, etc. Note that practically in all real flow cytometry datasets there exist many boundary points. This is mainly caused by some technical features of flow cytometry devices. As for cell debris removal, there exists a rich medical-based literature: since this topic is outside the scope of this book, the reader is invited to consult e.g. [176] and the references therein.

Figure 7.12 shows flow cytometry sample dataset (see its short description above, *PatientID* = *pid291*). The boundary points in the left panel are easily noticeable (located in the right and upper borders), whereas in the right panel they have been removed.

Another problem that should be addressed (solved or at least contained) is the heavy skewness in data. To reduce skewness in flow cytometry, certain transformations are used. The first popular transformation is the *inverse hyperbolic sine transformation* defined as

$$f(x) = \operatorname{asinh}(a + bx + c), \tag{7.30}$$

where usually, by default a and b are both equal to 1 and c to 0. The second popular transformation is the *biexponential transformation* (overparametrized inverse of the hyperbolic sine) defined as

$$f(x) = a\mathrm{e}^{b(x-w)} - c\mathrm{e}^{-d(x-w)} + g, \tag{7.31}$$

with default parameters selected to correspond to the hyperbolic sine, that is $a = 0.5$, $b = 1$, $c = 0.5$, $d = 1$, $g = 0$, $w = 0$. The two transformations give roughly the same results in practice.
Figure 7.13 shows the scatterplot of the original (not transformed) flow cytometry sample dataset (see its short description above, *PatientID=pid291*), while Fig. 7.14 shows the same dataset with the biexponential transformation employed. It is easily seen that this transformation has a noticeable positive effect.

Step 2: Standardize all variables to have zero mean and unit standard deviation. This step is motivated by the fact that in the *curvHDR* package a single parameter bandwidth matrix is used (i.e. $\mathbf{H} = h^2\mathbf{I}$). Note however, that the standardization may not always be sufficient, see a similar problem presented in Fig. 3.11.

Step 3: Identify feature significant regions as described in [51, 55] and briefly presented in Sect. 7.6.2.

Step 4: Replace every significant region obtained in Step 3 by a corresponding convex hull.

Step 5: Grow each convex hull by an empirical factor $G = 2^d$ (for explanation of such a particular value, see [125]).

Step 6: For each of the newly enlarged convex hull S, determine the subset of the data points lying inside that convex hull.

Step 7: For each points in every region S, calculate a kernel density estimate. In principle, any multivariate bandwidth selection method can be used to that end, see Chap. 4. The preferred method is the plug-in one, see Sect. 4.3.4, because it is probably the most robust and seems to be a popular choice by many authors). The FFT-based implementation presented in Chap. 5 is a clearly preferred option, as typical flow cytometry datasets are very big in size (many thousands of individual measurements).

Step 8: Select the searched gates based on kernel density estimates obtained in Step 7. This is done by taking a contour of the density function f in such a way that it covers a selected 'probability mass' of f. This is very difficult to give a default value for this parameter as it may vary a lot depending on a given flow cytometry dataset. A good starting reasonable value can be 90%.

Step 9: Determine the indices of the dataset points corresponding to the discovered gates.

Step 10: Transform the gates and gated data back to the original units.

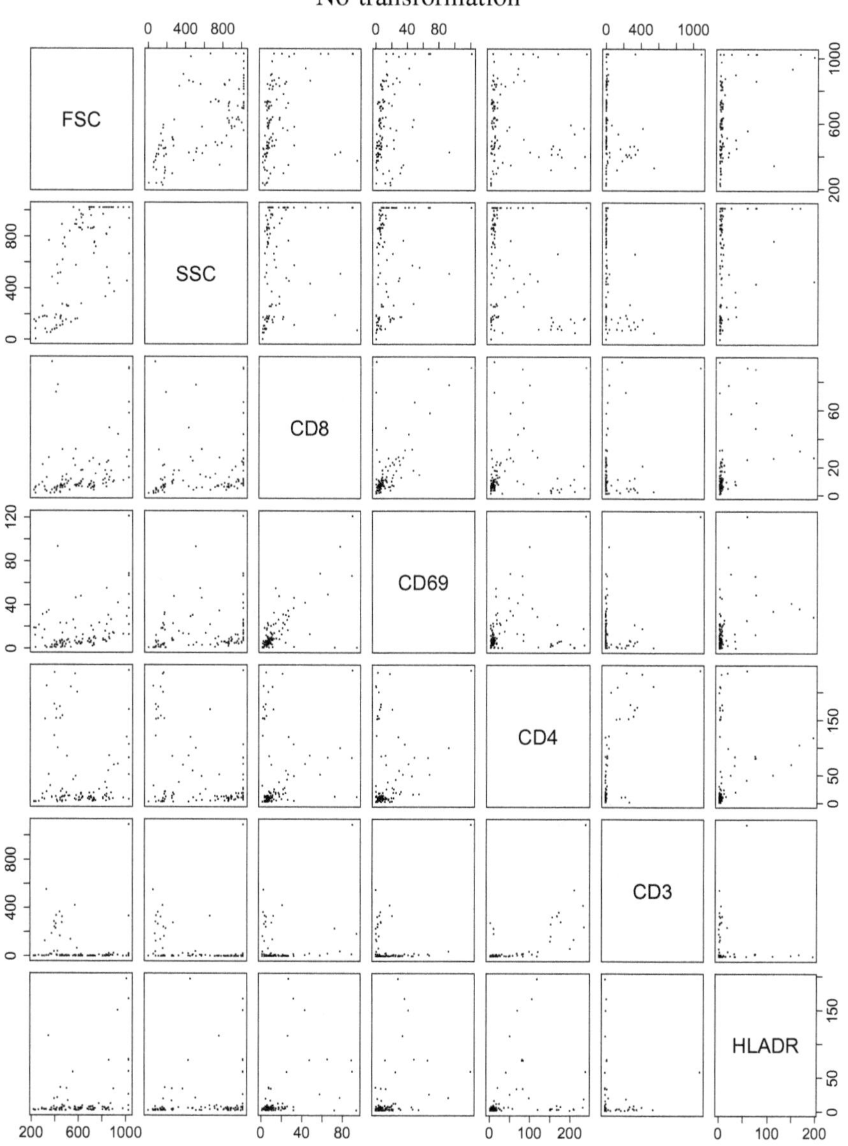

Fig. 7.13 The scatterplot of the original (untransformed) flow cytometry dataset

Figure 7.15 shows the main steps (from Step 3 to Step 8) of the gating algorithm. Note that in this example there was no need to perform data transformation (see Step 1), as the scatter plot of the FSC and SSC variables does not show any skewness. The only preprocessing performed was related to data standardization (see Step 2).

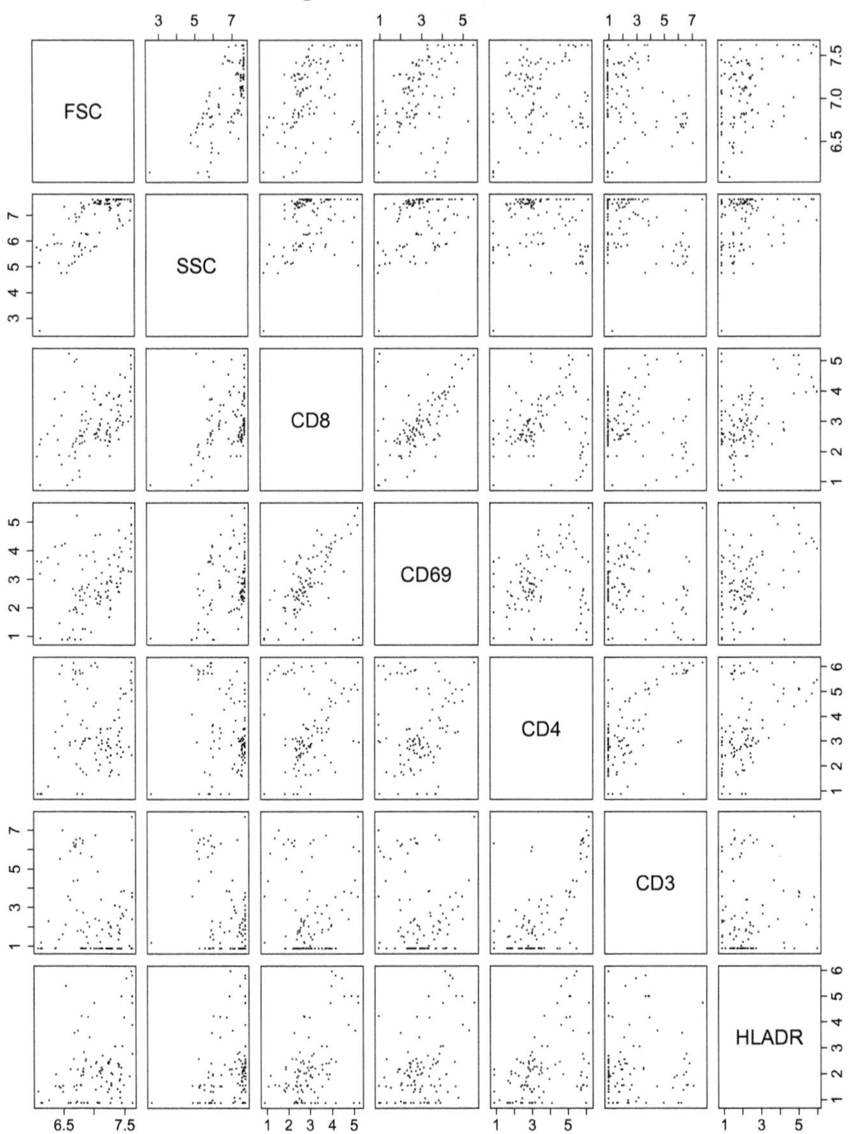

Fig. 7.14 The scatterplot of the flow cytometry dataset after the biexponential transformation

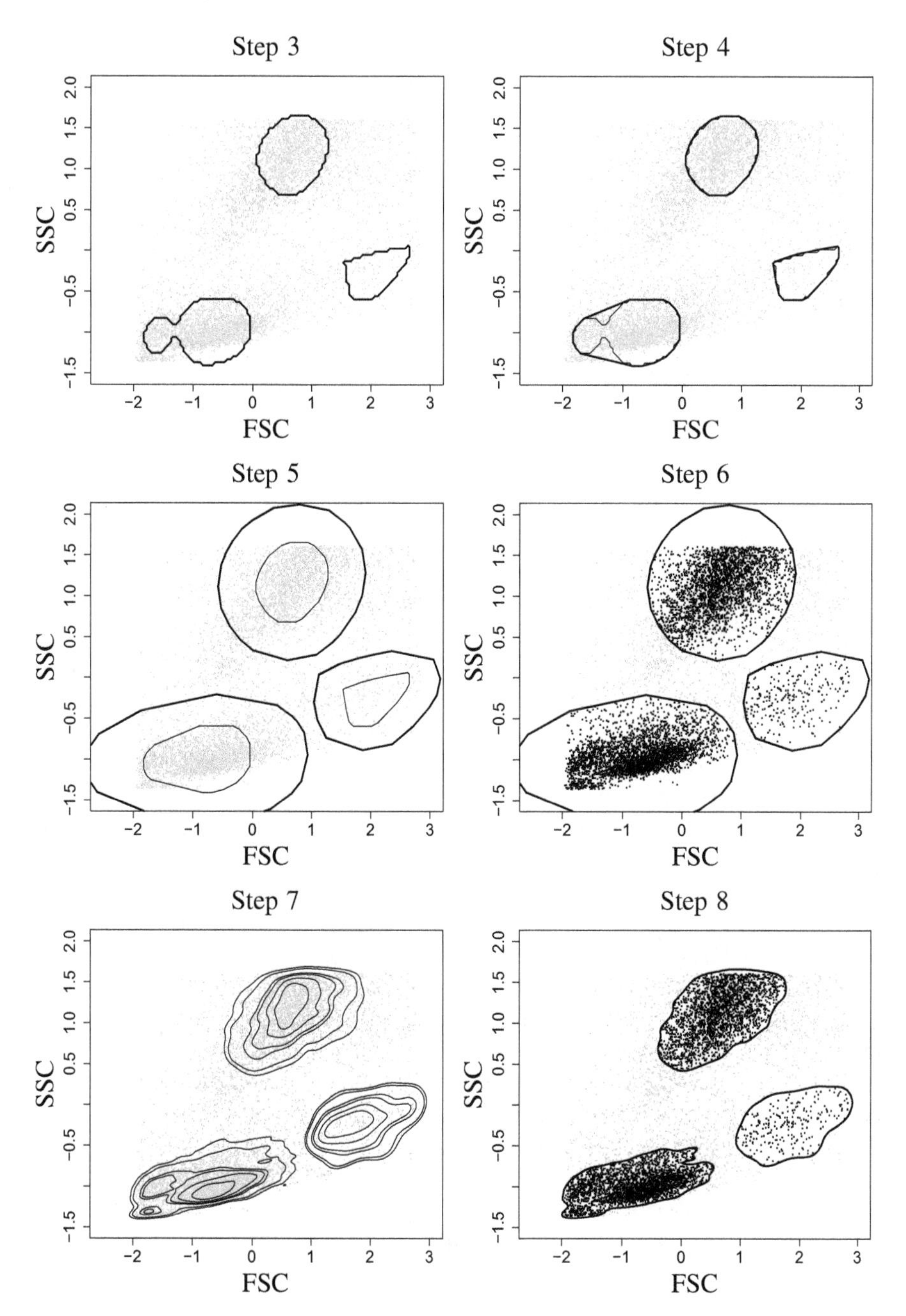

Fig. 7.15 The intermediate results of the gating algorithm

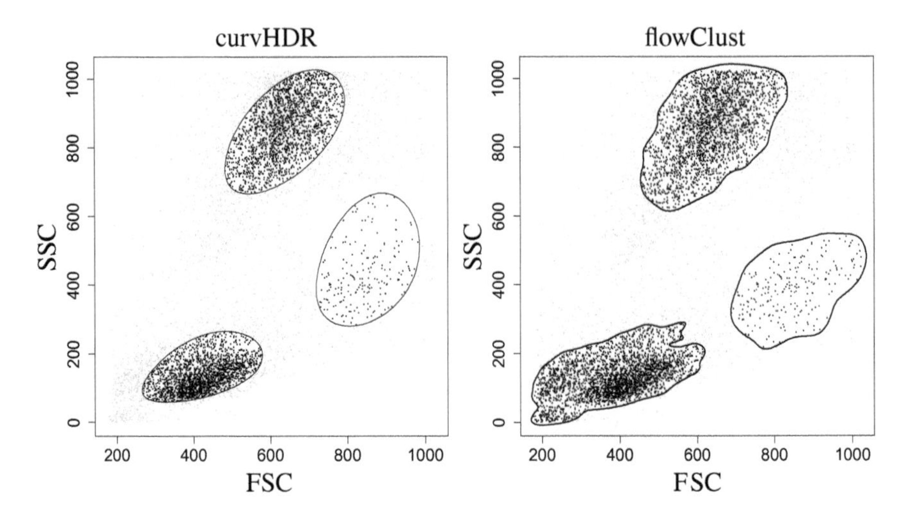

Fig. 7.16 A simple comparison of the gating results obtained using the *flowClust* package and the nonparametric method presented in this section

Steps 9 and 10 are not visualized, as these are very simple and require only certain dexterity in symbol manipulation.

Finally, note that the Bioconductor portal [10] contains a dedicated *R* package *flowClust* for robust model-based clustering using a t-mixture model with Box-Cox transformation. This package is parametric and hence differs from the one presented above. The methods employed in the *curvHDR* package are purely nonparametric and the resulted gate shapes have no restrictions. The main difference between the two methods is that *flowClust* requires specification of the number of clusters in advance and the gates discovered are always ellipses. Figure 7.16 shows a simple comparison of the two methods. The results are similar, but it seems that non-elliptical gates, to some degree, better reflect the spread of the data points. Obviously, the gating process as a whole is a very subjective task and the results presented here should be in a final step consulted with the flow cytometry experts.

It is also worth stressing that both methods are not fully automatic and the results depend strongly on many parameters that can be adjusted (see Table 1 in [125] and the help files contained in the *flowClust* package).

As the final point, it is worth noting that the *curvHDR* package was slightly modified by us to support unconstrained bandwidth matrices (in Step 7 above, see also Chap. 5). The original version of the *curvHDR* package uses diagonal bandwidth plug-in selector, because at the time of writing [125] the results presented in [72] had not been known.

Chapter 8
Conclusion and Further Research Directions

The kernel-based methods can be considered very effective smoothing techniques with many practical applications. They have been a point of focus for many researchers and the analysis of these methods has been an ongoing research task for many years. A number of publications and monographs that describe these have achieved a status of a classic reference.

It seems that the theory of kernel density estimation (and other problems related to KDE) has reached a kind of maturity and is very well established and understood. However, there is not that much progress in terms of performance improvement. It is especially important when nonparametric techniques are applied to large datasets and in cases, where the calculation time is a critical factor. Many authors either omit these problems or present them in a shallow way. This book tries to remedy this. It is primarily focused on the problem of estimation of the probability density but extending this methodology for other statistical characteristics (such as cumulative distribution function, receiver operating characteristic curves, survival function etc.) is not difficult.

The book is aimed at those who are new to the topics covered as well as those coming from a broader background related to nonparametric techniques.

Chapters 2 and 3 present certain fundamental concepts that are shown in practice on many suggestive examples based both on toy and real datasets.

Chapters 4 and 5 are intended for a more experienced reader with the latter being almost exclusively devoted to presenting the computational problems related to KDE and bandwidth selection. In these chapters, one also finds a number of numerical examples.

The subject matter of Chap. 6 is not something usually described in similar settings and it relates to author's original research results related to the problem of fast computation of a selected bandwidth algorithm based on utilizing Field-Programmable Gate Arrays (FPGA).

A. Gramacki, *Nonparametric Kernel Density Estimation and Its Computational Aspects*, Studies in Big Data 37,
https://doi.org/10.1007/978-3-319-71688-6_8

Chapter 7 presents five selected applications related to KDE—the first three are quite typical and are described by many authors (these topics being discriminant analysis, cluster analysis and kernel regression). The last two examples are more sophisticated. The penultimate example shows how KDE methodology can be used in the multivariate statistical process control area. Finally, the last example is related to analyzing the flow cytometry datasets (note that such datasets are, almost by definition, very big).

The following is a concise summary of the author's own contributions described in this book in relation to the most recent research in nonparametric smoothing methods:

- We have pointed out certain limitations of the FFT-based algorithm in terms of a fast computation of kernel density estimates. This is a very important issue, as many authors simply refer to this algorithm mechanically without describing potential pitfalls, including its greatest limitation.
- We have proposed a solution of the above-mentioned problem. After the improvements, the FFT-based algorithm is much more useful in practice as it supports the unconstrained bandwidth matrices and is not limited to their diagonal counterparts.
- We have developed a complete framework for using the FFT-based algorithm in terms of solving the bandwidth selection problem when the unconstrained bandwidths are involved. The method can be used for practically all state-of-the-art bandwidth selectors and its very good performance was confirmed by extensive numerical simulations. Moreover, the density derivatives of any order are also supported as well as integrated density derivative functionals.
- We have shown the practical usability of the current FPGA chips for solving pure numerical-like problems. Typically, FPGAs are used mainly for solving problems related to digital electronics and in similar contexts. Hence, our work can be treated as opening new avenue of research in terms of using FPGA devices in various nonparametric-like numerical problems.

Needless to say, notwithstanding the above-mentioned developments, there still remains a number of open problems regarding nonparametric smoothing. What follows is a short description of the research areas that seem promising in terms of further investigation.

- In certain KDE-related applications, such as the mean shift algorithm (see Sect. 7.3), the FFT-based algorithm presented in Chapter sec-FFT-based cannot be used since the evaluation of the density gradient for the mean shift algorithm is not performed over a grid of points but over sequences of iteratively shifted points starting at the sample data. The Improved Fast Gauss Transform method seems to be a good starting point here [17].
- Some more research is needed in terms of solving problems with stability of computations in cases, when the FFT-based method is used for bandwidth selection. This problem was pointed out in [72].

- The use of FPGA devices in solving pure numerical problems seems to be a very interesting avenue of research. In [73] the authors presented some preliminary results in that respect, showing the potential for further more in-depth study of these problems.

Bibliography

1. M. Abramowitz, I.A. Stegun, *Handbook of Mathematical Functions*. (Dover Publications, 1972), https://en.wikipedia.org/wiki/Abramowitz_and_Stegun (provides links to the full text which is in public domain)
2. B.K. Aldershof, Estimation of integrated squared density derivatives. Ph.D. thesis, A dissertation submitted to the faculty of The University of North Carolina at Chapel Hill, 1991
3. N. Altman, Ch. Léger, Bandwidth selection for kernel distribution function estimation. J. Stat. Plan. Inference. **46**(2), 195–214, 1995. https://doi.org/10.1016/0378-3758(94)00102-2
4. E. Anderson, The Irises of the Gaspé Peninsula. Bull. Am. Iris Soc. **59**, 2–5 (1935)
5. W. Andrzejewski, A. Gramacki, J. Gramacki, Graphics processing units in acceleration of bandwidth selection for kernel density estimation. Int. J. Appl. Math. Comput. Sci. **23**(4), 869–885 2013. https://doi.org/10.2478/amcs-2013-0065
6. A. Azzalini, A.W. Bowman, A look at some data on the old faithful geyser. J. R. Stat. Soc. Ser. C (Appl. Stat.) **39**(3), 357–365 (1990), http://www.jstor.org/stable/2347385
7. J. Bachrach, H. Vo, B. Richards, Y. Lee, A. Waterman, R. Avižienis, J. Wawrzynek, K. Asanović, Chisel: constructing hardware in a scala embedded language in *Design Automation Conference*, (IEEE, 2012), pp. 1212–1221, https://chisel.eecs.berkeley.edu
8. R.E. Bellman, *Dynamic programming*, (Princeton University Press, 1957)
9. P.K. Bhattacharya, Estimation of a probability density function and its derivatives. Indian J. Stat. Ser. A **29**, 373–382 (1967)
10. Bioconductor, *The open source software for bioinformatics*, http://www.bioconductor.org
11. Z.I. Botev, J.F. Grotowski, D.P. Kroese, Kernel density estimation via diffusion. Ann. Stat. **38**(5), 2916–2957 (2010). https://doi.org/10.1214/10-AOS799
12. A.W. Bowman, An alternative method of cross-validation for the smoothing of density estimates. Biometrika. **71**(2), 353–360 (1984), http://www.jstor.org/stable/2336252
13. A.W. Bowman, A. Azzalini, *Applied Smoothing Techniques for Data Analysis: The Kernel Approach with S-Plus Illustrations* (Oxford University Press, Oxford Statistical Science Series, 1997)
14. L. Breiman, W. Meisel, E. Purcell, Variable kernel estimates of multivariate densities. *Technometrics*. **19**(2), 135–144 (1977), http://www.jstor.org/stable/1268623
15. R. Cao, A. Cuevas, W.G. Manteiga, A comparative study of several smoothing methods in density estimation. Comput. Stat. Data Anal. **17**, 153–176 (1994). https://doi.org/10.1016/0167-9473(92)00066-Z
16. Y. Cao, H. He, H. Man, SOMKE: kernel density estimation over data streams by sequences of self-organizing maps. IEEE Trans. Neural Netw. Learn. Syst. **23**, 1254–1268 (2012). https://doi.org/10.1109/TNNLS.2012.2201167
17. J.E. Chacón, Personal communication, http://matematicas.unex.es/~jechacon/

A. Gramacki, *Nonparametric Kernel Density Estimation and Its Computational Aspects*, Studies in Big Data 37,
https://doi.org/10.1007/978-3-319-71688-6

18. J.E. Chacón, Data-driven choice of the smoothing parametrization for kernel density estimators. Can. J. Stat. **37**, 249–265 (2009). https://doi.org/10.1002/cjs.10016
19. J.E. Chacón, T. Duong, Multivariate plug-in bandwidth selection with unconstrained pilot bandwidth matrices. Test. **19**, 375–398 (2010). https://doi.org/10.1007/s11749-009-0168-4
20. J.E. Chacón, T. Duong, Unconstrained pilot selectors for smoothed cross validation. Aust. N. Z. J. Stat. **53**, 331–351 (2012). https://doi.org/10.1111/j.1467-842X.2011.00639.x
21. J.E. Chacón, T. Duong, Data-driven density derivative estimation, with applications to nonparametric clustering and bump hunting. Electron. J. Stat. **7**, 499–532 (2013). https://doi.org/10.1214/13-EJS781
22. J.E. Chacón, T. Duong, Efficient recursive algorithms for functionals based on higher order derivatives of the multivariate Gaussian density. Stat. Comput. **25**, 959–974 (2015). http://dx.doi.org/10.1007/s11222-014-9465-1
23. J.E. Chacón, T. Duong, M.P. Wand Asymptotics for general multivariate kernel density derivative estimators. Statistica Sinica **21**(2), 807–840 (2011), http://www.jstor.org/stable/24309542
24. J.E. Chacón, C. Tenreiro, Data-based choice of the number of pilot stages for plug-in bandwidth selection. Commun. Stat. Theory Methods. **42**(12), 2200–2214 (2013). http://dx.doi.org/10.1080/03610926.2011.606486
25. J.E. Chacóon, P. Monfort, A comparison of bandwidth selectors for mean shift clustering (2013). arXiv.org preprint
26. M. Charytanowicz, J. Niewczas, P. Kulczycki, P.A. Kowalski, S. Łukasik, S. Żak, in *Complete Gradient Clustering Algorithm for Features Analysis of X-Ray Images*, (Springer, Berlin, 2010), pp. 15–24. http://dx.doi.org/10.1007/978-3-642-13105-9_2
27. P. Chaudhuri, J.S. Marron, SiZer for exploration of structures in curves. J. Am. Stat. Assoc. **94**(447), 807–823 (1999), http://www.jstor.org/stable/2669996
28. Y. Cheng, Means shift, mode seeking, and clustering. IEEE Trans. Pattern Anal. Mach. Intell. **17**(8), 790–799 (1995). https://doi.org/10.1109/34.400568
29. S.-T. Chiu, Why bandwidth selectors tend to choose smaller bandwidths, and a remedy. Biometrika. **77**(1), 222–226 (1990), http://www.jstor.org/stable/2336068
30. S.-T. Chiu. The effect of discretization error on bandwidth selection for kernel density estimation. Biometrika. **78**(2), 436–441 (1991), http://www.jstor.org/stable/2337272
31. Y.-M. Chou, R.L. Mason, J.C. Young. The control chart for individual observations from a multivariate non-normal distribution. Commun. Stat. Theory Methods. **30**, 1937–1949 (2001). http://dx.doi.org/10.1081/STA-100105706
32. P.P. Chu, *FPGA Prototyping by VHDL Examples: Xilinx Spartan-3 Version* (Wiley Interscience, 2008)
33. D.B. Cline, J.D. Hart, Kernel estimation of densities with discontinuities or discontinuous derivatives. Statistics. **22**, 69–84 (1991). http://dx.doi.org/10.1080/02331889108802286
34. D. Comaniciu, P. Meer, Mean shift: a robust approach toward feature space analysis. IEEE Trans. Pattern Anal. Mach. Intell. **24**, 603–619 (2002). https://doi.org/10.1109/34.1000236
35. D. Comaniciu, V. Ramesh, P. Meer, Kernel-Based object tracking. IEEE Trans. Pattern Anal. Mach. Intell. **25**(5), 564–575 (2003). https://doi.org/10.1109/TPAMI.2003.1195991
36. P. Coussy, A. Morawiec. *High-Level Synthesis from Algorithm to Digital Circuit*. (Springer, Berlin, 2008). https://doi.org/0.1007/978-1-4020-8588-8
37. J. Ćwik, J. Mielniczuk. *Statystyczne systemy uczące się. Ćwiczenia w oparciu o pakiet R (Statistical Learning Systems. R Project based exercises)*. Oficyna Wydawnicza Politechniki Warszawskiej, 2005. (in Polish)
38. N. Daili, A. Guesmia, Remez algorithm applied to the best uniform polynomial approximations. Gen. Math. Notes **17**(1), 16–31 (2013)
39. L. Devroye, L. Györfi, *Nonparametric Density Estimation: The L1 View* (Wiley, New York, 1985)
40. L. Devroye, L. Györfi, A. Krzyżak, The Hilbert kernel regression estimate. J. Multivar. Anal. **65**, 209–227 1998. https://doi.org/10.1006/jmva.1997.1725

41. L. Devroye, A. Krzyżak, On the Hilbert kernel density estimate. Stat. Probab. Lett. **44**, 299–308 (1999). https://doi.org/10.1016/S0167-7152(99)00021-8
42. L. Devroye, G. Lugosi, *Combinatorial Methods in Density Estimation* (Springer, New York, 2001)
43. B. Droge, Some comments on cross-validation. Technical Report 1994–1997 (Humboldt Universitaet Berlin, 1996)
44. R.O. Duda, P.E. Hart, D.G. Stork, *Pattern Classification*, 2nd edn. (Wiley-Interscience, 2000)
45. T. Duong, Personal communication, http://www.mvstat.net/tduong/
46. T. Duong, Bandwidth selectors for multivariate kernel density estimation. Ph.D. thesis, University of Western Australia, School of Mathematics and Statistics, 2004
47. T. Duong, ks: Kernel density estimation and Kernel discriminant analysis for multivariate data in R. J. Stat. Softw. **21**, 1–16 (2007). http://dx.doi.org/10.18637/jss.v021.i07
48. T. Duong, Non-parametric smoothed estimation of multivariate cumulative distribution and survival functions, and receiver operating characteristic curves. J. Korean Stat. Soc. **45**, 33–50 (2016). https://doi.org/10.1016/j.jkss.2015.06.002
49. T. Duong, *Kernel Smoothing*, (R package version 1.10.6, 2017), http://CRAN.R-project.org/package=ks
50. T. Duong, G. Beck, H. Azzag, M. Lebbah, Nearest neighbour estimators of density derivatives, with application to mean shift clustering. Pattern Recognit. Lett. **80**, 224–230 (2016). https://doi.org/10.1016/j.patrec.2016.06.021
51. T. Duong, A. Cowling, I. Koch, M.P. Wand, Feature significance for multivariate kernel density estimation. Comput. Stat. Data Anal. **52**, 4225–4242 (2008). https://doi.org/10.1016/j.csda.2008.02.035
52. T. Duong, M.L. Hazelton, Plug-in bandwidth matrices for bivariate kernel density estimation. J. Nonparametric Stat. **15**, 17–30 (2003). http://dx.doi.org/10.1080/10485250306039
53. T. Duong, M.L. Hazelton, Convergence rates for unconstrained bandwidth matrix selectors in multivariate kernel density estimation. J. Multivar. Anal. **93**, 417–433 (2005). https://doi.org/10.1016/j.jmva.2004.04.004
54. T. Duong, M.L. Hazelton, Cross-validation Bandwidth Matrices for Multivariate Kernel Density Estimation. Scand. J. Stat. **32**, 485–506 (2005). https://doi.org/10.1111/j.1467-9469.2005.00445.x
55. T. Duong, M. Wand. *Feature: Local Inferential Feature Significance for Multivariate Kernel Density Estimation*, (R package version 1.2.13, 2015), http://CRAN.R-project.org/package=feature
56. V.A. Epanechnikov, Nonparametric estimation of a multidimensional probability density. Theory Probab. Appl. **14**, 153–158 (1969). http://dx.doi.org/10.1137/1114019
57. M. Ester, H.-P. Kriegel, J. Sander, X. Xu, A density-based algorithm for discovering clusters in large spatial databases with noise, in *Proceedings of 2nd International Conference on Knowledge Discovery and Data Mining (KDD-96)*, (1996), pp. 226–231
58. B. S. Everitt, S. Landau, M. Leese, D. Stahl. *Cluster Analysis*, 5th edn. (Wiley Series in Probability and Statistics, 2011)
59. B.S. Everitt, A. Skrondal, *The Cambridge Dictionary of Statistics*, 4th edn. (Cambridge University Press, 2010)
60. W. Feller, *An Introduction to Probability Theory and Its Applications*, (Wiley, 1968)
61. R. A. Fisher. The use of multiple measurements in taxonomic problems. Ann. Eugen. **7**, 179–188 (1936). http://dx.doi.org/10.1111/j.1469-1809.1936.tb02137.x
62. E. Fix, J.L. Hodges, *Discriminatory Analysis - Nonparametric Discrimination: Consistency Properties*, (Technical report USAF School of Aviation Medicine, Randolph Field, Texas, 1951)
63. K. Fukunaga, L. Hostetler, The estimation of the gradient of a density function, with applications in pattern recognition. IEEE Trans. Inf. Theory. **21**, 32–40 (1975). https://doi.org/10.1109/TIT.1975.1055330
64. Ch.R. Genovese, M. Perone-Pacifico, I. Verdinelli, L. Wasserman, On the path density of a gradient field. Ann. Stat. **37**(6A), 3236–3271 (2009). https://doi.org/10.1214/08-AOS671

65. A.K. Ghost, Optimal smoothing in kernel discriminant analysis. Comput. Stat. Data Anal. **50**, 3113–3123 (2006). https://doi.org/10.1016/j.csda.2005.06.007
66. A.K. Ghost, P. Chaudhuri, Optimal smoothing in kernel discriminant analysis. Statistica Sinica. **14**(2), 457–483 (2004), http://www.jstor.org/stable/24307204
67. A.L. Givan, *Flow Cytometry: First Principles* (Wiley-Liss, New York, 2001)
68. F. Godtliebsen, J.S. Marron, P. Chaudhuri, Significance in scale space for bivariate density estimation. J. Comput. Graph. Stat. **11**, 1–21 (2002). http://dx.doi.org/10.1198/106186002317375596
69. W. González-Manteiga, C. Sánchez-Sellero, M.P. Wand, Accuracy of binned kernel functional approximations. Comput. Stat. Data Anal. **22**, 1–16 (1996). https://doi.org/10.1016/0167-9473(96)88030-3
70. N. Gopalakrishnan, *Flow cytometry data analysis using Bioconductor—A typical work flow*, 2009. The paper can be downloaded from https://www.bioconductor.org/help/workflows/highthroughputassays/flowWorkFlow.pdf, the data files can be downloaded from www.bioconductor.org/help/workflows/high-throughput-assays/dataFiles.tar
71. A. Gramacki, J. Gramacki, FFT-Based fast computation of multivariate kernel estimators with unconstrained bandwidth matrices. J. Comput. Graph. Stat. **26**, 459–462 (2016). http://dx.doi.org/10.1080/10618600.2016.1182918
72. A. Gramacki, J. Gramacki, FFT-based fast bandwidth selector for multivariate kernel density estimation. Comput. Stat. Data Anal. **106**, 27–45 (2017). https://doi.org/10.1016/j.csda.2016.09.001
73. A. Gramacki, M. Sawerwain, J. Gramacki, FPGA-based bandwidth selection for kernel density estimation using high level synthesis approach. Bull. Pol. Acad. Sci. Tech. Sci. **64** (2016). https://doi.org/10.1515/bpasts-2016-0091
74. L. Greengard,J. Strain, The fast gauss transform. SIAM J. Sci. Stat. Comput. **12**, 79–94 (1991). http://dx.doi.org/10.1137/0912004
75. L. Györfi, M. Kohler, A. Krzyżak, H. Walk, *A Distribution-free Theory of Nonparametric Regression*, (Springer, 2002)
76. P. Hall, K.-H. Kang, Bandwidth choice for nonparametric classification. Ann. Stat. **33**, 284–306 (2005). https://doi.org/10.1214/009053604000000959
77. P. Hall, J.S. Marron, B.U. Park, Smoothed cross validation. Probab. Theory Relat. Fields. **92**, 1–20 (1992). https://doi.org/10.1007/BF01205233
78. P. Hall, J.S. Marron, Estimation of integrated squared density derivatives. Stat. Probab. Lett. **6**(2), 109–115 (1987). https://doi.org/10.1016/0167-7152(87)90083-6
79. P. Hall, B.U. Park, New methods for bias correction at endpoints and boundaries. Ann. Stat. **30**(5), 1460–1479 (2002), http://www.jstor.org/stable/1558721
80. P. Hall, M.P. Wand, On the accuracy of binned kernel density estimators. J. Multivar. Anal. **56**(2), 165–184 (1996). https://doi.org/10.1006/jmva.1996.0009
81. W. Härdle, *Applied Nonparametric Regression, Econometric society monographs*, (Cambridge University Press, Cambridge, 1990)
82. W. Härdle, *Smoothing Techniques: With Implementation in S*, (Springer Series in Statistics, 1991)
83. W. Härdle, J.S. Marron, M.P. Wand, Bandwidth choice for density derivatives. J. R. Stat. Soc. Ser. B (Methodological). **52**(1), 223–232 (1990), https://www.jstor.org/stable/2345661
84. W. Härdle, M. Müller, S. Sperlich, A. Werwatz, *Nonparametric and Semiparametric Models*, (Springer Series in Statistics, 2004)
85. W. Härdle, D.W. Scott, Smoothing in low and high dimensions by weighted averaging using rounded points. Comput. Stat. **7**, 97–128 (1992)
86. J.A. Hartigan, M.A. Wong, Algorithm AS 136: A K-means clustering algorithm. J. R. Stat. Soc. Ser. C (Appl. Stat.) **28**(1), 100–108 (1979), http://www.jstor.org/stable/2346830
87. T. Hastie, R. Tibshirani, J. Friedman, *The Elements of Statistical Learning: Data Mining, Inference, and Prediction*, (Springer Series in Statistics, 2001), https://web.stanford.edu/~hastie/ElemStatLearn

88. T. Hayfield, J.S. Racine, The np package: kernel methods for categorical and continuous data. R News. **7**(2), 35–43 (2007), http://CRAN.R-project.org/doc/Rnews/Rnews_2007-2.pdf
89. N.-B. Heidenreich, A. Schindler, S. Sperlich, Bandwidth selection for kernel density estimation: a review of fully automatic selectors. AStA Adv. Stat. Anal. **97**, 403–433 (2013). https://doi.org/10.1007/s10182-013-0216-y
90. H. Heinz, Density estimation over data streams. Ph.D. thesis, Philipps-Universität Marburg, Faculty of Mathematics and Computer Science, 2007
91. H.V. Henderson, S.R. Searle, Vec and vech operators for matrices, with some uses in Jacobians and multivariate statistics. Can. J. Stat. **7**(1), 65–81 (1979), http://www.jstor.org/stable/3315017
92. N. Hjort, M. Jones, Locally parametric nonparametric density estimation. Ann. Stat. **24**(4), 1619–1647 (1996), http://www.jstor.org/stable/2242742
93. B. Holmquist, The d-variate vector hermite polynomial of order k. Linear Algebr. Appl. **237–238**, 155–190 (1996). https://doi.org/10.1016/0024-3795(95)00595-1
94. I. Horová, J. Koláček, J. Zelinka, *Kernel Smoothing in Matlab: Theory and Practice of Kernel Smoothing* (World Scientific Publishing, Singapore, 2012)
95. H. Hotelling, The generalization of student's ratio. Ann. Math. Stat. **2**(3), 360–378 (1931). https://doi.org/10.1214/aoms/1177732979
96. O. Hryniewicz, *On the Robustness of the Shewhart Control Chart to Different Types of Dependencies in Data*, (Springer, Berlin, 2012), pp. 19–33. https://doi.org/10.1007/978-3-7908-2846-7_2
97. A.J. Izenman, *Modern Multivariate Statistical Techniques*, (Springer Texts in Statistics, 2008)
98. W. Jarosz, Efficient Monte Carlo methods for light transport in scattering media. Ph.D. thesis, UC San Diego, Sep 2008, https://cs.dartmouth.edu/~wjarosz/publications/dissertation
99. M.C. Jones, Variable kernel density estimates and variable kernel density estimates. Aust. N. Z. J. Stat. **32**, 361–371 (1990). https://doi.org/10.1111/j.1467-842X.1990.tb01031.x
100. M.C. Jones, Simple boundary correction for kernel density estimation. Stat. Comput. **3**(3), 135–146 (1993). https://doi.org/10.1007/BF00147776
101. M.C. Jones, On kernel density derivative estimation. Commun. Stat. Theory Methods. **23**(8), 2133–2139 (1994). http://dx.doi.org/10.1080/03610929408831377
102. M.C. Jones, P.J. Foster, Generalized jackknifing and higher order kernels. J. Nonparametric Stat. **3**(1), 81–94 (1993). http://dx.doi.org/10.1080/10485259308832573
103. M.C. Jones, J.S. Marron, S.J. Sheather, A brief survey of bandwidth selection for density estimation. J. Am. Stat. Assoc. **91**(433), 401–407 (1996). http://dx.doi.org/10.1080/01621459.1996.10476701
104. M.C. Jones, J.S. Marron, S.J. Sheather, Progress in data-based bandwidth selection for kernel density estimation. Comput. Stat. **11**(3), 337–381 (1996), http://oro.open.ac.uk/id/eprint/24974
105. J. Koronacki, J. Ćwik. *Statystyczne systemy uczące się (Statistical Learning Systems)*. Wydawnictwo Naukowo-Techniczne, 2005. (in Polish)
106. HP. Kriegel, P. Kröger, J. Sander, A. Zimek, Density-based clustering. WIREs Data Min. Knowl. Discov. **1**, 231–240 (2011). https://doi.org/10.1002/widm.30
107. P. Kulczycki. *Estymatory jądrowe w analizie systemowej (Kernel Estimators in Systems Analysis)*. Wydawnictwo Naukowo-Techniczne, 2005. (in Polish)
108. P. Kulczycki, M. Charytanowicz, A complete gradient clustering algorithm formed with kernel estimators. Int. J. Appl. Math. Comput. Sci. **20**(1), 123–134 (2010). https://doi.org/10.2478/v10006-010-0009-3
109. P. Kulczycki, M. Charytanowicz, A.L. Dawidowicz, A convenient ready-to-use algorithm for a conditional quantile estimator. Appl. Math. Inf. Sci. **9**(2), 841–850 (2015)
110. P.A. Kulczycki, A.L. Dawidowicz, Kernel estimator of quantile. *Universitatis Iagellonicae Acta Mathematica*. XXXVII:325–336 (1999)
111. D. Lang, M. Klaas, N. Freitas, Empirical testing of fast kernel density estimation algorithms. Technical Report UBC TR-2005-03 (University of British Columbia, Department of Computer Science, 2005), http://www.cs.ubc.ca/~nando/papers/empirical.pdf

112. C.E. Lawrence, *Partial Differential Equations*, (American Mathematical Society, 2010)
113. D.O. Loftsgaarden, C.P. Quesenberry, A nonparametric estimate of a multivariate density function. Ann. Math. Stat. **36**(3), 1049–1051 (1965), http://www.jstor.org/stable/2238216
114. S. Łukasik, Parallel computing of kernel density estimates with MPI, in *International Conference on Computational Science ICCS 2007*. LNCS, vol. 4489 (Springer, Berlin, 2007), pp. 726–733. https://doi.org/10.1007/978-3-540-72588-6_120
115. J.S. Marron, D. Nolan, Canonical kernels for density estimation. Stat. Probab. Lett. **7**, 195–199 (1988). https://doi.org/10.1016/0167-7152(88)90050-8
116. J.S. Marron, D Ruppert, Transformations to reduce boundary bias in kernel density estimation. J. R. Stat. Soc. Ser. B (Methodol.) **56**, 653–671 (1994), http://www.jstor.org/stable/2346189
117. R.L. Mason, J.C. Young, ASA-SIAM series on statistics and applied mathematics, in *Multivariate Statistical Process Control with Industrial Applications* (2002)
118. J. Matai, D. Richmond, D. Lee, R. Kastner, Enabling FPGAs for the Masses (2014), arXiv:1408.5870
119. G.J. McLachlan, *Discriminant Analysis and Statistical Pattern Recognition*, (Wiley Interscience, 2004)
120. G. Menardi, *Variable Kernel Density Estimate: A New Proposal*, (Cleup, Padova, 2006), pp. 579–582
121. D.C. Montgomery, *Introduction to Statistical Quality Control*, 5th edn. (Wiley, 2005)
122. H.-G. Müller, Smooth optimum kernel estimators near endpoints. Biometrika **78**, 521–530 (1991), http://www.jstor.org/stable/2337021
123. B.J. Murphy, M.A. Moran, Parametric and kernel density methods in discriminant analysis: Another comparison. Comput. Math. Appl. **12**, 197–207 (1986). https://doi.org/10.1016/0898-1221(86)90073-8
124. E.A. Nadaraya, On estimating regression. Theory Probab. Appl. **9**, 141–142 (1964). http://dx.doi.org/10.1137/1109020
125. U. Naumann, G. Luta, M.P. Wand, The curvHDR method for gating flow cytometry samples. BMC Bioinf. **11**(44), 1–13 (2010). https://doi.org/10.1186/1471-2105-11-44
126. R.B. Nelsen. *An Introduction to Copulas*. Springer Series in Statistics (2006)
127. M.G. Ormerod, *Flow Cytometry—A Basic Introduction* (2008), http://flowbook.denovosoftware.com
128. S. Palnitkar, *Verilog HDL (paperback)*, 2nd edn. (Prentice Hall, 2003)
129. B.U. Park, J.S. Marron, Comparison of data-driven bandwidth selectors. J. Am. Stat. Assoc. **85**, 66–72 (1990). https://doi.org/10.1080/01621459.1990.10475307
130. E. Parzen, On estimation of a probability density function and mode. Ann. Math. Stat. **33**(3), 1065–1076 (1962). https://doi.org/10.1214/aoms/1177704472
131. L.E. Peterson, K-nearest neighbor, *Scholarpedia the peer-reviewed open-access encyclopedia*, **4**(2), 1883 (2009). http://dx.doi.org/10.4249/scholarpedia.1883
132. P. Phaladiganon, S.B. Kim, V.C.P. Chen, J.-G. Baek, S.-K. Park, Bootstrap-Based T^2 multivariate control charts. Commun. Stat. Simul. Comput. **40**, 645–662 (2011). http://dx.doi.org/10.1080/03610918.2010.549989
133. PLUGIN, The plug-in source codes, 2016, https://github.com/qMSUZ/plugin
134. B.L.S. PrakasaRao, *Nonparametric Functional Estimation*, (Academic Press, Orlando, 1983)
135. R Core Team, *R: A Language and Environment for Statistical Computing*, (R Foundation for Statistical Computing, Vienna, Austria, 2016), https://www.R-project.org/
136. R Core Team, *Introduction to Flow Cytometry: A Learning Guide*, Manual Part number: 11-11032-01 rev. 2, (R Foundation for Statistical Computing, San Jose, California, USA, 2002), https://www.bdbiosciences.com/us/support/s/itf_launch
137. J.S. Racine, T. Hayfield, *NP: Nonparametric Kernel Smoothing Methods for Mixed Data Types*, (R package version 0.60-3, 2016), https://CRAN.R-project.org/package=np
138. E. Rafajłowicz, Nonparametric orthogonal series estimators of regression: a class attaining the optimal convergence rate in L_2. Stat. Probab. Lett. **5**, 219–224 (1987). https://doi.org/10.1016/0167-7152(87)90044-7

139. E. Rafajłowicz, Nonparametric least squares estimation of a regression function. Stat. A J. Theor. Appl. Stat. **19**, 349–358 (1988). http://dx.doi.org/10.1080/02331888808802107
140. E. Rafajłowicz, E. Skubalska-Rafajłowicz, FFT in calculating nonparametric regression estimate based on trigonometric series. Appl. Math. Comput. Sci. **3**(4), 713–720 (1993)
141. E. Rafajłowicz, E. Skubalska-Rafajłowicz, Nonparametric regression estimation by Bernstein-Durrmeyer polynomials. Tatra Mt. Math. Publ. **17**, 227–239 (1999)
142. V. C. Raykar, R. Duraiswami. Fast optimal bandwidth selection for kernel density estimation. In Society for Industrial and Applied Mathematics, editors, *Proceedings of the 2006 SIAM International Conference on Data Mining*, (2006), pp. 524–528
143. V.C. Raykar, R. Duraiswami, Very fast optimal bandwidth selection for univariate kernel density estimation. Technical Report CS-TR-4774/UMIACS-TR-2005-73 (Department of Computer Science, University of Maryland, College Park, 2006)
144. V.C. Raykar, R. Duraiswami, L.H. Zhao, Fast computation of kernel estimators. J. Comput. Graph. Stat. **19**(1), 205–220 (2010), http://www.jstor.org/stable/25651308
145. E.Ya. Remez, Sur la détermination des polynômes d'approximation de degré donnée. Comm. Soc. Math. Kharkov **10**, 41–63 (1934)
146. M. Rosenblatt, Remarks on some nonparametric estimates of a density function. Ann. Math. Stat. **27**(3), 832–837 (1956), https://projecteuclid.org/euclid.aoms/1177728190
147. M. Rudemo, Empirical choice of histograms and kernel density estimators. Scand. J. Stat. **9**(2), 65–78 (1982), http://www.jstor.org/stable/4615859
148. S.R. Sain, Adaptive kernel density estimation. Ph.D. thesis, Rice University, Huston, USA, 1994
149. S.R. Sain, Multivariate locally adaptive density estimation. Comput. Stat. Data Anal. **39**, 165–186 (2002). https://doi.org/10.1016/S0167-9473(01)00053-6
150. S.R. Sain, K.A. Baggerly, D.W. Scott, Cross-Validation of multivariate densities. J. Am. Stat. Assoc. **89**(427), 807–817 (1994), http://www.jstor.org/stable/2290906
151. I.H. Salgado-Ugarte, M.A. Pérez-Hernández, Exploring the use of variable bandwidth kernel density estimators. Stata J. **3**(2), 133–147 (2003)
152. K. Schauer, T. Duong, C.S. Gomes-Santos, B. Goud, Studying intracellular trafficking pathways with probabilistic density maps. Methods Cell Biol. **118** 325–343 (2013). https://doi.org/10.1016/B978-0-12-417164-0.00020-3
153. W.R. Schucany, Locally optimal window widths for kernel density estimation with large samples. Stat. Probab. Lett. **7**, 401–405 (1989). https://doi.org/10.1016/0167-7152(89)90094-1
154. E.F. Schuster, Estimation of a probability density function and its derivatives. Ann. Math. Stat. **40**(4), 1187–1195 (1969)
155. D. W. Scott. *Using Computer-Binned Data For Density Estimation*, (Springer, New York, 1981), pp. 292–294. https://doi.org/10.1007/978-1-4613-9464-8_42
156. D.W. Scott, Averaged shifted histograms: effective nonparametric density estimators in several dimensions. The Ann. Stat. **13**(3), 1024–1040 (1985), http://www.jstor.org/stable/2241123
157. D.W. Scott, Feasibility of multivariate density estimates. Biometrika. **78**, 197–205 (1991). https://doi.org/10.1093/biomet/78.1.197
158. D.W. Scott, *Multivariate Density Estimation: Theory, Practice, and Visualization*, (Wiley, 1992)
159. D.W. Scott, Averaged shifted histogram. WIREs Comput. Stat. **2**, 160–164 (2010). https://doi.org/10.1002/wics.54
160. D.W. Scott, *Multivariate Density Estimation and Visualization*, (Springer, Berlin, 2012), pp. 549–569
161. D.W. Scott, S.J. Sheather, Kernel density estimation with binned data. Commun. Stat. Theory Methods. **14**, 1353–1359 (1985). https://doi.org/10.1080/03610928508828980
162. D.W. Scott, G.R. Terrell, Biased and unbiased cross-validation in density estimation. J. Am. Stat. Assoc. **82**(400), 1131–1146 (1987), http://www.jstor.org/stable/2289391
163. L.Scrucca. qcc: an r package for quality control charting and statistical process control. R News. 4/1, 11–17 (2004), https://cran.r-project.org/doc/Rnews/

164. H.M. Shapiro, *Practical Flow Cytometry*, (Wiley, 2003)
165. S.J. Sheather, M.C. Jones, A reliable data-based bandwidth selection method for kernel density estimation. J. R. Stat. Soc. Ser. B (Methodolo.) **53**(3), 683–690 (1991)
166. W.A. Shewhart, Economic quality control of manufactured product. Bell Syst. Tech. J. **9**, 364–389 (1930). https://doi.org/10.1002/j.1538-7305.1930.tb00373.x
167. B.W. Silverman, Algorithm AS 176: kernel density estimation using the fast fourier transform. J. R. Stat. Soc. Ser. C (Appl. Stat.) **31**(1), 93–99 (1982), http://www.jstor.org/stable/2347084
168. B.W. Silverman, *Density Estimation for Statistics and Data Analysis* (Chapman and Hall/CRC, London, 1986)
169. J.S. Simonoff, Smoothing categorical data. J. Stat. Plan. Inference. **47**(1–2), 41–69 (1995). https://doi.org/10.1016/0378-3758(94)00121-B
170. J.S. Simonoff, *Smoothing Methods in Statistics* (Springer, New York, 1996)
171. E. Skubalska-Rafajłowicz, Random projection RBF nets for multidimensional density estimation. Int. J. Appl. Math. Comput. Sci. **18**(4), 455–464 (2008). http://dx.doi.org/10.2478/v10006-008-0040-9
172. E. Skubalska-Rafajłowicz, Random projections and hotelling's T^2 statistics for change detection in high-dimensional data streams. Int. J. Appl. Math. Comput. Sci. **23**(2), 447–461 (2013). http://dx.doi.org/10.2478/amcs-2013-0034
173. S.W. Smith, *The Scientist and Engineer's Guide to Digital Signal Processing*, (California Technical Publishing, 1997), http://www.dspguide.com
174. C.O.S. Sorzano, J. Vargas, A. Pascual Montano, A survey of dimensionality reduction techniques. (2014) arXiv.org preprint, https://arxiv.org/abs/1403.2877
175. Synflow, Synflow Cx, (2017), https://www.synflow.com. Accessed 25 Apr 2017
176. P. Terho, O.Lassila, Novel method for cell debris removal in the flow cytometric cell cycle analysis using carboxy-fluorescein diacetate succinimidyl ester. Cytom. Part A. **69A**, 552–554 (2006). https://doi.org/10.1002/cyto.a.20261
177. G.R Terrell, The maximal smoothing principle in density estimation. J. Am. Stat. Assoc. **85**(410), 470–477 (1990), http://www.jstor.org/stable/2289786
178. G.R. Terrell, D.W. Scott, Variable kernel density estimation. Ann. Stat. **20**(3), 1236–1265 (1992), http://www.jstor.org/stable/2242011
179. J.R. Thompson, R.A. Tapia, *Nonparametric Function Estimation, Modeling, and Simulation*, (SIAM, 1990). https://doi.org/10.1137/1.9781611971712
180. H. Tristen, J.S. Racine, Nonparametric econometrics: the np package. J. Stat. Softw. **27**, 1–32 (2008). http://dx.doi.org/10.18637/jss.v027.i05
181. B.A. Turlach, *Bandwidth Selection in Kernel Density Estimation: A Review Report*, (Université Catholique de Louvain, Belgium, C.O.R.E. and Institut de Statistique, 1993)
182. UNICEF, *The state of the Word's children 2003*, (Oxford University Press, For UNICEF, 2013), https://www.unicef.org/sowc03
183. C.P. Verschoor, A. Lelic, J.L. Bramson, D.M.E. Bowdish, An introduction to automated flow cytometry gating tools and their implementation. Front. Immunol. **6**(380) (2015). https://doi.org/10.3389/fimmu.2015.00380
184. J.E. Volder, The CORDIC trigonometric computing technique. IRE Trans. Electron. Comput. **EC-8**, 330–334 (1959). https://doi.org/10.1109/TEC.1959.5222693
185. J.S. Walther, A unified algorithm for elementary functions, in *Proceedings of Spring Joint Computer Conference*, (1971), pp. 379–385. https://doi.org/10.1145/1478786.1478840
186. M.P. Wand, Fast computation of multivariate kernel estimators. J. Comput. Graph. Stat. **3**(4), 433–445 (1994), http://www.jstor.org/stable/1390904
187. M.P. Wand, M.C. Jones, Multivariate plug-in bandwidth selection. Comput. Stat. **9**(2), 97–116 (1994)
188. M.P. Wand, M.C. Jones, *Kernel Smoothing*, (Chapman and Hall, 1995)
189. M.P. Wand, G. Luta, U. Naumann, *curvHDR: Filtering of Flow Cytometry Samples*, (R package version 1.1-0, 2016), https://CRAN.R-project.org/package=curvHDR
190. M.P. Wand, J.S. Marron, D. Ruppert, Transformations in density estimation. J. Am. Stat. Assoc. **86**(414), 343–353 (1991), http://www.jstor.org/stable/2290569

191. M.P. Wand, B. Ripley, *Functions for Kernel Smoothing Supporting Wand and Jones (1995)*, (R package version 2.23-15, 2015), http://CRAN.R-project.org/package=KernSmooth
192. D. Wang, O. Hryniewicz, A fuzzy nonparametric Shewhart chart based on the bootstrap approach. Int. J. Appl. Math. Comput. Sci. **25**(2), 389–401 (2015). https://doi.org/10.1515/amcs-2015-0030
193. G.S. Watson, Smooth regression analysis. Sankhyā: Indian J. Stat. Ser. A **26**(4), 359–372 (1964), http://www.jstor.org/stable/25049340
194. G.B. Wetherill, D.W. Brown, *Statistical Process Control: Theory and Practice*, (Chapman and Hall/CRC Texts in Statistical Science, 1991)
195. J. Yan, *Multivariate Modeling with Copulas and Engineering Applications*, (Springer, London, 2006), pp. 973–990. https://doi.org/10.1007/978-1-84628-288-1_51
196. M.J. Zaki, W. Meira, *Data Mining and Analysis: Fundamental Concepts and Algorithms*, (Cambridge University Press, 2014), http://www.dataminingbook.info/pmwiki.php/Main/BookDownload (provides links to the full text which is for personal online use)
197. S. Zhang, R. Karunamuni, M.C. Jones, An improved estimator of the density function at the boundary. J. Am. Stat. Assoc. **94**(448), 1231–1241 (1999), http://www.jstor.org/stable/2669937
198. S. Zhang, R.J. Karunamuni, On kernel density estimation near endpoints. J. Stat. Plan. Inference. **70**, 301–316 (1998). https://doi.org/10.1016/S0378-3758(97)00187-0
199. S. Zhang, R.J. Karunamuni, On nonparametric density estimation at the boundary. J. Nonparametric Stat. **12**, 197–221 (2000). http://dx.doi.org/10.1080/10485250008832805
200. A. Zhou, Z. Cai, L. Wei, W. Qian, M-kernel merging: towards density estimation over data streams, in *Eighth International Conference on Database Systems for Advanced Applications (DASFAA 2003)*, (2003) pp. 1–9
201. M. Zwolinski, *Digital System Design with VHDL*, (Prentice Hall, 2004)
202. K.Żychaluk, P.N. Patil, A cross-validation method for data with ties in kernel density estimation. Ann. Inst. Stat. Math. **60**, 21–44 (2008). https://doi.org/10.1007/s10463-006-0077-1

Index

A. Gramacki, *Nonparametric Kernel Density Estimation and Its Computational Aspects*, Studies in Big Data 37,
https://doi.org/10.1007/978-3-319-71688-6